Digital Transformations for Lighting in the Workplace

Digital Transformations for Lighting in the Workplace: A Systematic Approach Used in Ergonomics offers a practical concept for the implementation of digital transformation in ergonomics in work settings. It specifically focuses on providing information about illumination in production and non-production fields, and the described design solutions are applicable in practice. The concepts can be used in a typical manufacturing workplace or an academic setting.

The methods in the book complement the modern trend to digitalize the workplace, making the research and practical outcomes of this book compatible with the concept of Industry 4.0 – Digital Manufacturing. This title offers a systematic approach to the field of digital transformation for ergonomics. It presents an opportunity for the reader to learn to create a digital model for lighting by analyzing mathematical models for calculation through formulas and simulation algorithms. To put learning into context, this book provides two case studies from the production and non-production sectors, including an example of a classroom. The reader will then be able to utilize the methods to create their own digitized illumination system.

This monograph is an ideal read for academics and researchers working at universities in the field of Ergonomics and professionals in industrial management including those in manufacturing plants, ergonomists, designers from the industry sector, or people who are interested in ergonomics, digitization, and simulation of a working environment.

Digital Transformations for Lighting in the Workplace

A Systematic Approach Used in Ergonomics

Darina Dupláková and Ján Duplák

CRC Press is an imprint of the
Taylor & Francis Group, an informa business

First edition published 2023
by CRC Press
6000 Broken Sound Parkway NW, Suite 300, Boca Raton, FL 33487-2742

and by CRC Press
4 Park Square, Milton Park, Abingdon, Oxon, OX14 4RN

CRC Press is an imprint of Taylor & Francis Group, LLC

Library of Congress Cataloging-in-Publication Data
Names: Dupláková, Darina, author. | Duplák, Ján, author.
Title: Digital transformations for lighting in the workplace : a systematic approach used in ergonomics / Darina Dupláková, Ján Duplák.
Description: First edition. | Boca Raton : CRC Press, 2023. | Includes bibliographical references and index.
Identifiers: LCCN 2022038101 (print) | LCCN 2022038102 (ebook) | ISBN 9781032412542 (hardback) | ISBN 9781032418308 (paperback) | ISBN 9781003359937 (ebook)
Subjects: LCSH: Office buildings--Lighting--Computer simulation. | Factories--Lighting--Computer simulation.
Classification: LCC TK4399.O35 D87 2023 (print) | LCC TK4399.O35 (ebook) | DDC 621.32/2523011--dc23/eng/20221017
LC record available at https://lccn.loc.gov/2022038101
LC ebook record available at https://lccn.loc.gov/2022038102

ISBN: 978-1-032-41254-2 (hbk)
ISBN: 978-1-032-41830-8 (pbk)
ISBN: 978-1-003-35993-7 (ebk)

DOI: 10.1201/9781003359937

Typeset in Times
by MPS Limited, Dehradun

Review process
This scientific monograph has been reviewed by two reviewers and approved by Publishing Committee as a part of the publication process.

Acknowledgement
The scientific monograph was created with the support of the VEGA research grant 1/0431/21 Research of light-technical parameters in production hall using digital ergonomics tools.

Contents

Preface

The present times are increasingly influenced by digital technologies. Advancement of these technologies with the purpose of comprehensive improvement of closely personalized chain in society, in synergy with its increasing efficiency, allows for greater connectivity of sectors and branches in the area of services, production, system solutions, etc. Current modern principles of digitalization represent the concept of creating an integrated PC system, which consists of models, simulations, analyses, as well as 3D visualizations. These tools assist each other in creating new solutions, a dynamic environment with the ability to develop continuous innovation and rationalization measures. A comprehensive integrated system that can be created by using the elements of digitalization is gradually shaped also in the field of ergonomics, most often in ergonomic assessment and design.

The present scientific monograph describes the issue of a systematic approach to digital transformation in ergonomics with the focus on lighting in the work environment. This issue is elaborated on gradually in five main chapters. The first three chapters are primarily devoted to theoretical knowledge and its development in the field, specifically focusing on the Scientific Paradigm of Ergonomics, Ergonomics and lighting in workplace, and Digitalization and Digital transformation in ergonomics. In the fourth chapter, the authors describe practical results of their own research aimed at implementing digital ergonomic transformation in the engineering industry. The practical implementation of their own research is also presented in the final chapter, which provides information on the implementation of digital transformation in ergonomic design with a focus on non-production space.

The scientific monograph was created with the support of the VEGA research grant 1/0431/21 Research of light-technical parameters in production hall using digital ergonomics tools.

Preface

Acknowledgement

The scientific monograph was created with the support of the VEGA research grant 1/0431/21 Research of light-technical parameters in production hall using digital ergonomics tools.

Author Biographies

Darina Dupláková is Associate Professor in the Faculty of Manufacturing Technologies at the Technical University in Košice, Slovakia. Her research interests include ergonomic rationalization, digital ergonomics, lighting simulation, data analysis, production process simulation and planning. In 2019, she was awarded as "Personality of Science and Technology under 35 Years", for excellent scientific results in the field of lighting technology by the International Society of Engineering Pedagogues.

Ján Duplák is Associate Professor in the Faculty of Manufacturing Technologies at the Technical University of Kosice, Slovakia. His research interests include advanced technologies, cutting materials, CNC machining and programming, simulation, and manufacturing.

Scientific Paradigm of Ergonomics

1

Knowledge of ergonomic principles, general principles, and methodology is now becoming a necessity in every area, whether it is the manufacturing or non-production space. In the manufacturing sector, many products are added every day, which requires the creation of new work processes arising in new workplaces, which need to be either created or redesigned, given their condition and nature. In these activities, it is necessary to implement ergonomic principles and procedures specified exactly for the research area. For the needs of practical development of this scientific monograph, the first chapter is prepared dealing with a basic overview of general ergonomics and summaries of the theoretical approach to workplace designing in ergonomics.

1.1 BRIEF OVERVIEW OF GENERAL ERGONOMICS

In a dynamically developing modern society, ergonomics can be considered an interdisciplinary scientific field with an indispensable place in science, research, and practice. The most common are systems and subsystems of ergonomics implemented in the area of humanities, technical, and medical disciplines. This scientific field represents an interaction between people, machines, and factors influencing this interaction. The essence of ergonomics lies in its very name, which is the union of two Greek words – Ergon, which means work, and nomos, which means the laws of nature. The very essence can be considered as the basis on which each of the hitherto interpreted significant definitions of this concept is based.

DOI: 10.1201/9781003359937-1

In the "Industrial Health and Safety" encyclopaedia, the concept of ergonomics is interpreted along two planes. One indicates the area of scientific and technical knowledge in relation to a person's work, and the other shows indicators of how this knowledge is used to achieve a higher level of mutual adaptation of a person to their work from a humane (health) and economic (labour productivity) point of view.

According to the International Labour Organization, the term ergonomics is defined as follows:

> *Humanization of work, achieving a higher level of adaptation between man and his work from the humane (health) and economic (labour productivity) point of view. According to the authors, the subject matter of ergonomics is the study of interactions in mainly work systems, discovery of their mutual links and effects, and creation of sets of measures of a technical, organizational, and personnel type, such as the application of relevant knowledge to design of the means of work, to furnishing and arrangement of work places, to creation of a healthy work environment, to creation of an appropriate mode and organization of work and to the preparation of man to competently carry out the anticipated work, etc.*

Based on ISO/CD6385, ergonomics is understood as

> *a comprehensive integration of knowledge from the life sciences ensuring interconnection between products, work system, activities performed, and the work environment with mental and physical limits and abilities of people. Within this complex definition, it seeks to ensure health, safety and well-being in synergy with the rationalisation of performance and efficiency of employees' work activities.*

Last but not least, it is necessary to mention the definition of ergonomics also according to the international ergonomic organization (IEA), which is as follows:

> *Ergonomics is a scientific discipline based on understanding the interactions between man and other components in the system. By applying appropriate methods, theory, and data, it improves human health, well-being, and performance. It contributes to addressing the design and evaluation of work, tasks, products, environment, and system to be compatible with people's needs, capabilities, and performance constraints. Ergonomics is, therefore, a systemically oriented discipline that covers virtually all aspects of human activity. As part of a holistic approach, it includes physical, cognitive, social, organisational, environmental and other relevant factors.*

Based on the above ergonomic guidelines, the basic subject matter of ergonomic research can be summarized and compared in developmental stages of interdisciplinarity of this scientific field. The following paragraph provides a comparison of the development of the subject matter under investigation. The sequence in which the subject of the examination by ergonomics has been developing can be divided as follows:

Initial status:

- Exploring the human-machine relationship, human-work environment
- Examination of human limit values
- Means of transport, production technology
- Corrective ergonomics
- Comprehensive ergonomic studies
- Ergonomic rationalization, design
- Low level of publishing

Current status:

- Exploring the human-work relationship
- Optimization of human activity
- Production and non-production workplaces
- Designing workplaces
- CAD, 3D, ergonomic software
- Simulations, digital enterprise
- Educational activity deepens

Summary of the subject matter of ergonomic research creates the basis for determining the goal of ergonomics, which can be considered to be maintaining health, i.e., physical, mental, and social satisfaction of a person, creating conditions for optimal human activity, as well as creating a sense of well-being in the workplace. Another objective may also be understood to be optimising the mental and physical load of a person by means of a comprehensive solution to the problem raised, from partial ties to the search for a suitable solution by using the knowledge of the aforementioned related scientific disciplines. In its complexity, the primary objective thus defined determines the classification of ergonomics in two basic directions that contribute to ergonomic solutions from the perspective of the present or the future. These are the following bases of ergonomics classification:

- *Corrective ergonomics* – creates space for ergonomic assessment of the existing human-machine-environment system, the results bring proposals for measures ensuring increased efficiency of the complex system.

- *Preventive ergonomics* – deals with future human activity and the conditions in which the activity will be carried out, the implementation phase is included in the stage of planning and designing work systems.

In modern ergonomics, at the level of current knowledge and technical possibilities of modern science, the original natural procedure has been applied to tools development until the advent of the scientific and technical revolution, namely practical verification of the solution from the impact on the health of employees all the way to economic benefits. Ergonomics is further developed in theoretical and applied fields. It integrates applied parts of all scientific disciplines that will allow increased efficiency of human work while not damaging employee health and, at the same time, achieving economic benefits. Modern ergonomics is currently classified on the basis of conditions in practice focusing on two sub-areas: micro-ergonomics and macro-ergonomics.

Micro-ergonomics focuses on solving everyday problems of an enterprise and increasing the efficiency of human labour. It is characterized by the application of operational approaches. The principles of participatory ergonomics are applied here. Here, human labour is considered to be effective when employees are able to perform work tasks in such a way that qualitative and quantitative production requirements are met without harming their health. In order to increase market competitiveness, the aim is to train or attract the most versatile staff and equip operations with flexible multi-purpose equipment so that the enterprise has the widest possible range of options to respond to the state of supply and demand on the market. In these conditions, ergonomics is no longer only a question of humanity and ethics, but also becomes the subject of economic intentions.

Micro-ergonomics implements enterprise ergonomic programs that were created in the environment of a market economy, where they allow enterprises to achieve competitiveness and long-term stability. They are aimed at protecting workers' health and economic benefits. The measures taken shall be evaluated in the form of impact studies. Health impacts allow for the evaluation of methodologies for the epidemiology of non-infectious diseases and the economic benefits are assessed by cost-benefit analyses (CBAs).

Practically proven micro-ergonomic knowledge is generalized within the macro-ergonomic framework, where it is scientifically reassessed and summarized in databases and programmes, which then serve as a basis for enterprises, but also form the basis for development of TUR (sustainable development) strategies on a global scale (that is to say, a micro-solution to macro problems).

Macro-ergonomics applies mainly proactive approaches, collects into databases, and integrates the knowledge of all scientific disciplines and its own basic and applied research, the application of which will facilitate and streamline human work while achieving economic benefits. It creates databases and models for global use. These databases are used in the process of developing national legislation and strategies, in the design of new work and organizational systems, but also in their rationalization within micro-ergonomics. It is implemented either in scientific institutions or in larger advanced enterprises, where sufficient number of micro-ergonomic findings emerges in the application of ergonomic programmes in operations.

In addition to the above classification of modern ergonomics, ergonomics can be categorized into three main areas based on the specification of the International Ergonomics Association (IEA) as follows:

- Physical ergonomics – deals with the impact of working conditions and work environment on human health. It provides options for analysing the process of receiving information, the ability to process and apply it based on the acquired data, knowledge, and previous experience. These include, for example, the issue of work positions, objects handling, repeated work activities, work-induced diseases, especially of the locomotive system, workplace arrangement, work safety.
- Cognitive (mental) ergonomics – focuses on the psychological aspects of work. This includes mental load, decision-making processes, skills.
- Organizational ergonomics – focuses on the optimization of social and technical systems, including their organizational structures, strategies, procedures, etc.

The main areas of categorization are continuously complemented by the IEA with specific areas, including the following:

- Myoskeletal ergonomics – deals with prevention of work-induced diseases of the musculoskeletal system, especially diseases of the spine and upper limbs resulting from overload. These diseases have a gradual onset and the relative risk of acquiring them is increased by ergonomic exposure. For example, forced position or repetition of movements can be considered ergonomic exposure. The risk is relative, because factors not related to work, such as inflammatory processes in the body or metabolic disorders, may also be involved in development of these diseases.
- Psychosocial ergonomics deals with psychological demands at work and stress factors. The psychological requirements of work and the degree of decision-making affect the level of stress of the worker in dealing with the work situation. This area is closely

related to myoskeletal ergonomics as stress and other psychosocial factors influence the frequency and severity of musculoskeletal disorders.
- Participatory ergonomics – draws on the premise of participation of employees themselves in the design and implementation of changes at the workplace. Employees' active role leads to a better understanding of their difficulties and motivates possible ergonomic adjustments of the workplace and its conditions.
- Rehabilitation ergonomics focuses on people with disabilities, their professional training, and technical measures of the work environment.

Individual ergonomic levels, partial attributes, and parameters create a complex ergonomic system in synergy with each other, which is described as a system of three basic elements: man, machine, environment, interacting with each other in time. The basic interactions of the ergonomic system most often include the following ones:

- man – tool
- man – machine
- man – technical equipment
- man – environment
- machine – environment

An ergonomic system is a set of elements that are functionally connected to each other and allow the intended outputs – results – to be achieved from the given inputs. It is a dynamic open system, the specificity of which is the man.

Essential for an existing or emerging ergonomic system is implementation of the assessment on the basis of predetermined criteria. Basic criteria for the ergonomic system assessment shall be:

- productivity – the amount of labour performed per unit of time,
- reliability – timely and error-free completion of tasks; it is quantified by the probability of faultless operation,
- economy – determines financial costs (costs per unit of production, rate of return period, etc.),
- intensity of physical strain of the system function – we measure energy consumption per working cycle or per unit of time,
- intensity of mental strain of the system function – we determine the mental load that a given system induces in a person,
- hazard of the system – that is, the risk to health by injury – the opposite is occupational safety and health (OSH),
- level of the system's hygiene – eliminate the hazard of falling ill,

- aesthetics of the system – sensitivity to the beauty of individual elements of the system,
- ergonomics – a comprehensive approach to assessing the overall ergonomic level of machines,
- flexibility – the possibility of flexible changes of the system,
- stability – the tendency to keep variable values within given limits and maintain them there.

The existing ergonomic system and its attributes should be observable as synergies of the resulting concept on the following planes:

- economic – increased labour productivity, eliminated unproductive work, improved performance, reduced number, or absence of accidents – decreased costs of treatment, etc.,
- social – good interpersonal relations in the workplace, professional competitiveness, improvement proposals, etc.,
- health-related – reduced workload, good health of workers – physical fitness and the like.

Although current conditions require general ergonomics to be differentiated in a myriad of ways, each partial component generates, in synergy and coordination with the other attributes, a number of benefits, a summary of which is given in the text below.

General ergonomic benefits for organizations

- increased work performance
- reduced error rate and confusion
- reduced incidence of sick leave of employees and occupational diseases
- reduced fluctuation and absenteeism

General ergonomic benefits for workers

- improvement of the worker's mental and physical condition
- minimized symptoms of mental and physical fatigue
- improved self-actualization (benefits social sphere)
- positive impact on the economic situation of the individual and the family

General ergonomic benefits for society

- improved health of the population
- development of companies
- increased standard of living

1.2 MILESTONES OF ERGONOMICS – A BRIEF SUMMARY

The beginnings and roots of ergonomic approaches date back to the early stages of human development. It is not, of course, a concept and principles as we know them today. Since the very beginning of the historical development of humanity, a certain intermediary article is inserted between man and the subject of his work as part of his work activity, which is intended to improve the resulting effect of that work activity. These intermediate links are developing very intensely, not only co-determining the way of work, but also affecting the worker who uses them. The use of adequate tools is generally a strategy that contributes to the efficiency of the activities carried out. It is not only the domain of man, but it can also be found in animals, too. Thanks to the level of development of the central nervous system and its second signal system, man was able to gradually improve his tools on the basis of the knowledge he acquired from verifying those tools in practice, from the form of a stone fist wedge to the stone axe about 1.8 million years ago to complicated modern instruments, machines, and devices. The oldest artifacts found prove that already the predecessors of man worked with natural materials and adapted work tools to their individual needs. Throughout the history, we can observe that gradual specialization and division of labour has led to further improvement of tools, working methods, but also to development of machines. Craftsmen could adapt their tools, machines, working environment, organization of work individually, according to their physical size, strength, habits, and experience depending on the level of their intellect.

The process described above was disrupted by the advent of the scientific and industrial revolution. Development continued from artisanal to centralized production, from manufactures to factory mass production, which in fact gradually ended the period of spontaneous individual adaptation of labour and working conditions to man. Till present, custom-made production has survived since that period. The situation changed with the onset of the scientific and technological revolution in the 17th century and the subsequent "industrial revolution" with development of mass production in the first half of the 19th century, when production was centralized and concentrated in specialized enterprises and companies.

Just as seemingly unlimited natural resources began to be recklessly exploited, so too was human labour. During this period, the possibilities for individual verification and adaptation of work equipment to individual workers became more difficult. Machines and tools started to be produced en masse, the level of adaptation of new machines and equipment to the workers from a given population corresponded to the level of knowledge of their creators, the technicians.

However, it turned out that with his capabilities, man limits the reliability of the work system function. As a result of maladjusted work and working conditions, health, economic, and social consequences began to emerge. Although the emergence of ergonomics can be described quite well, the period of the birth of this new discipline was long and torturous, so its beginnings cannot be precisely dated. The initial increase in interest in the area of relations between man and his modern working environment can be noted in the period around the First World War. The workers in the factories with arms production in England were essential for sustaining the war effort, but there were numerous unexpected complications in the efforts to increase arms production. This led in 1915 to the creation of the "Health of Munitions Workers' Committee", employing persons trained in physiology and psychology. After the war, the commission was converted into the Industrial Fatigue Research Board. A wide range of problems was addressed individually or also jointly by trained psychologists, physiologists, medical doctors, and technicians. The outbreak of the Second World War brought rapid development in the field of military technology. The first official efforts in this area were already called "Engineering Psychology" and "Human Engineering" during the Second World War, which developed in the United Kingdom and the United States.

However, as an independent scientific discipline, ergonomics – focusing on the systemic solution of the entire complex of human problems at work – only emerged after the Second World War. Its creation is tied with 12 July 1949, when a meeting of the Admiralty was held in London, Great Britain, where the first interdisciplinary group was formed, which was interested in the issue of man at work. Later, at the meeting of this group on 16 February 1950 the name "ergonomics" was adopted for this new discipline.

For the first time, however, the term "ergonomics" was used by a Polish educator and scientist Wojciech Jastrzębowski already in 1857 in the debate "*Rys ERGONOMJI or NAUKI O PRACY, opartnej na pravdach poczerpniętych z Nauki Przyrody*" (Sketch of ergonomics or the science of work based on the truths derived from the natural sciences); reprint, Centralny Instytut Ochrony Pracy, Warzsawa, 1997. Jastrzębowski was a professor of natural sciences at the Agronomic Institute in Warsaw-Marymount and an expert in national physiography. The first meeting of the IEA was held in Stockholm, Sweden, in 1961. Currently, this association has active groups in most European countries in the USA, Japan, and also in Australia.

1.3 THEORETICAL APPROACH TO WORKPLACE DESIGNING IN ERGONOMICS

Applying ergonomics requires a broad understanding of this discipline. Ergonomics promotes a "holistic approach" characterized by the fact that it deals with complete and integrated systems rather than with their parts. This approach takes into account physical, cognitive, social, organizational, environmental, and all other relevant factors (Figure 1.1). Ergonomics professionals often work in specialized sectors or application areas. Areas of application are not mutually exclusive but are continuously developing.

In terms of ergonomics and its principles, the workplace constitutes a part of a larger whole – the work environment. The work environment is characterized by the sum of all material conditions of work activity related to other conditions, which is a set of all tangible and intangible factors directly affecting the worker and his work activity. It is actually the living environment at workplaces.

The concept of the work environment can also be defined as a simultaneous impact of external tangible and intangible factors affecting the worker and his work activity. More broadly, it is a set of conditions under which a production process is carried out through a variety of means and labour. In a narrower sense, this is an integral part of the production space in which workers carry out various types of work.

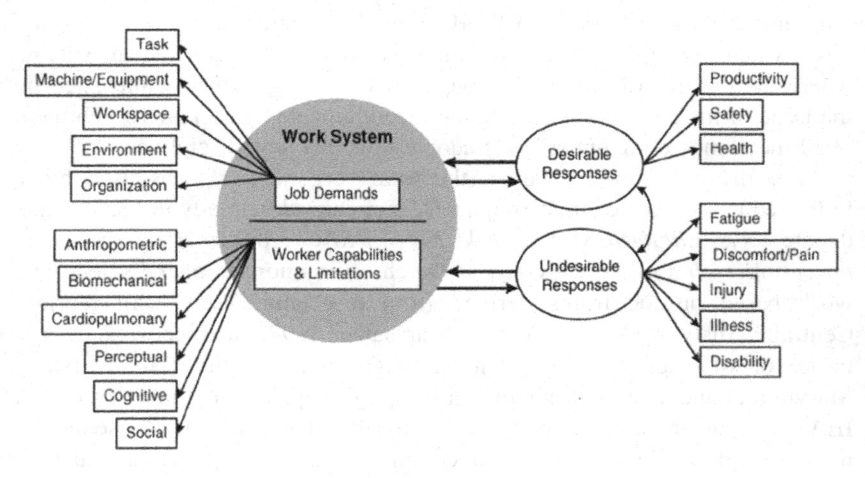

FIGURE 1.1 Interaction model of ergonomic work system.

The work environment has been one of the main areas of interest for early occupational psychologists (since about 1910), with a particular focus on the impact of climatic conditions (e.g., light, temperature, ventilation) on employee productivity. During these years of research, several important relationships have been discovered such as between excessive noise and employee health and productivity. However, the publication of Hawthorn's experiments, which after lengthy field research failed to prove the relationship between changes in the physical environment and employee productivity, resulted in a general decline in the scientific community's interest in studying the effects of the work environment until the 1960s.

Between 1940 and 1960, there were few studies on the working environment, but this topic was not completely neglected. Social psychologists and construction schools have constantly studied people's interactions with the environment (though not necessarily with the work environment) and demonstrated how manipulation of the physical environment can cause differences in how people interact with each other. For example, the spatial arrangement of furniture has been found to affect the number and nature of conversations between people, and the placement of people in a building helps determine who they interact with and make friends with.

After large-scale offices began to be built en masse in the United States between 1960 and 1970, occupational psychologists again began to address the relationship between workers and their work environment. As I mentioned at the beginning of the chapter, proponents of large-capacity offices saw the advantage of this type of workplace, especially in better communication between employees. This statement helped persuade a heap of corporations to start experimenting with demolishing the interior walls of the office, and so large-scale offices became a means of reducing their second largest operating cost (the first being human resources). Gradually, companies increased the density of office occupancy of offices and reduced the distance between individual desks, while neglecting the inclusion of plants in the office, the correct placement of desks and air circulation. Consequently, concerns were raised about the efficiency or effectiveness of these offices, which facilitated a new wave of research into large-capacity offices. This research focused on the identification of key aspects influencing the results and attitudes of employees and the description of the mechanisms of their functioning. These aspects include light, climatic conditions, privacy and noise, but also density, respective number of employees sharing one office.

Generally, if professionals want to evaluate the work environment as a whole, they must also take into account the synergy of all the critical factors that influence the work process and not just evaluate them selectively.

The methodology of comprehensive evaluation of the working environment consists of the following steps:

- Selection of the place where the evaluation will take place (workplace, job position).
- Selection of critical factors for the job position (questionnaire form).
- Evaluation of critical factors according to the importance of their impact on the work environment (Paired evaluation method – Fuller's triangle).
- Selection of the 10 most critical factors using the previous step.
- Weight distribution of factors depending on the effect on human health.
- Calculation of the ratio of average values of factors.
- Self-assessment based on the completion of a completed questionnaire (subjective assessment).
- Measurement of measurable critical factors (objective evaluation).
- Assignment of point evaluation by employees of the work environment to immeasurable factors (questionnaire – objective evaluation).
- Calculation of the factor standard for individual critical factors. (Normalization of indicators in the interval >0.1>).
- Completion of factors into the resulting summary table.
- Highlighting, identification of critical problem areas within the most critical factors of the workplace.
- Evaluation and assessment of measurements.
- Implementation of adjustments to the working environment to reduce the negative impact of the environment on workers.
- Regular monitoring of the work environment.

The objectives of the evaluation process are divided into:

- evaluation of critical factors of the working environment,
- evaluation of typical factors of the working environment,
- evaluation of prescribed factors of the working environment,
- evaluation of selected factors of the working environment,
- evaluation of factor categories,
- evaluation of the complex quality of the working environment.

Evaluation of critical factors of the work environment is a process in which they are evaluated, respective assessed factors that may be limiting or exceed critical values set by legislation and thus endanger human health, the state of the environment.

Evaluation of typical work environment factors – assessment of typical work environment factors and their measurement, is a process in which it is necessary to define e.g., goal and type of work environment, time during which the worker is in this work environment, etc. It is an evaluation of measured data, which are then processed, either absolutely (simple quantification) or relative (quantification of the state of the environment with respect to other factors).

The assessment of the prescribed factors of the working environment concerns the assessment of the factors prescribed by the legislation, which stipulates the number of measurements, the accessible values in a certain working environment and also the exact measurement methods.

The evaluation of selected factors of the working environment is carried out in case it is necessary to evaluate the state of one or less number of factors, which could occur negative phenomena, but in terms of quality they are seen as problem-free.

Evaluation of factor categories includes evaluations of specific acting factors included in individual classes. This assessment assesses the impact of a whole class of factors.

Evaluation of the complex quality of the working environment is the highest level of evaluation. It is typical for this evaluation that a larger number of factors is measured, which are subsequently evaluated from the absolute evaluation of the state of only critical factors to the relative evaluation of all factors affecting the environment.

The analysis of workplace factors is carried out in two ways – the qualitative and the quantitative. In qualitative analysis, the following activities are carried out: a comparative base is established along with evaluation criteria, and the comparative analysis is made. Quantitative analysis takes into account: duration of the parameter's action, parameter's influence, the number of parameters that act simultaneously, and the importance of the impact of the parameters evaluated.

The basic subjective evaluation methods may include:

- Distributing questionnaires – subjective evaluation method, the essence is to obtain employee feedback on job satisfaction in a given work environment, it is applied through a questionnaire with different scaling.
- Delphi method – a subjective method that is used to determine an expert estimate through a group of experts. It is used mainly to predict risks, more often in management, but it is also used to evaluate the work environment.
- Group assessment – an evaluation method based on the experience and knowledge of experts in the field, using mainly an analogy and a comparative approach.

The following shall be included among the basic methods of objective evaluation:

- points assignment method – analytical method of evaluation, determination of all risk factors acting on the work environment, a disadvantage of this method is its high subjectivity and inaccuracy arising from the selection of requirements and weights.
- work environment evaluation through coefficients – assessment based on coefficients of severity of impact of the load parameters Q_{dif}, which determines how much the real load is greater than the permissible load.
- mathematical methods – objective methods, using a combination of previous methods and statistics, different mathematical models are compiled (dynamic and static, linear, and non-linear, deterministic, and stochastic) due to gravity of different types of factors.

The basis for an objective and quantifiable evaluation of the work environment is the determination of the intensity and duration of the impact of parameters on the human body. A key attribute is also the correct selection and subsequent assessment of critical workplace factors or the work environment as a whole. For a comprehensive work environment evaluation and determination of the most critical factors impinging thereon, the pairwise assessment method is most often used in practice, which consists of determining 20 most serious factors adversely affecting the work environment with their subsequent ranking using the Fuller triangle.

The Fuller triangle is one of the evaluation options based on a comparison of always two criteria, out of which it is necessary to choose the more important one. A fixed numbering of the criterion in numerical order of 1, 2, …, n is a prerequisite. The Fuller triangle is formed by double lines in which each pair of criteria is located just once, subject to occurrence of each pair of criteria opposite each other just once. In each pair the evaluator then identifies the criterion that they consider to be more important. When comparing each of the two criteria out of a total of k criteria, the combinations of the two elements are selected from k. From the resulting number of comparisons, the values of the weights are then determined in accordance with the number of those marked i (n_i). Individual critical factors are then assigned normalized factors by means of conversion of the values measured in the interval <0; 1>.

According to the effect of the factor on humans and based on theoretical analyses, a weight factor can be assigned to individual effect indicators. The highest value of the factor in this interval is considered to be the factor with the greatest effect. The weights of indicator factors' effects are determined from the outputs of the pair assessment.

In the penultimate part of the process, the so-called vector of weight density is determined, obtained by converting the sum of the ratios of the average values of the factors per unit system.

The total work environment load is determined in conjunction with the exact expression of the actual load by individual factors according to the equation:

$$\lambda_X = H_{VVX} \cdot P_{TX} \tag{1.1}$$

where

λ_x – actual load by factor x,
H_{VVx} – vector of weight density for factor x,
P_{Tx} – standardized factor value,

whereby, based on the above equation, the total work environment load can be determined as follows:

$$\lambda_C = \sum_{i=1}^{n} \lambda_i \tag{1.2}$$

For a comprehensive evaluation of the working environment, it is inevitable to obtain data on objectification, quantitative and qualitative assessment, as well as data on the working environment itself and the workers performing work in the given environment. To supplement the missing, primarily subjective data, a so-called supplementary questionnaire, is aimed at collecting data directly from workers performing tasks in the assessed work environment.

However, no legislative or normative regulations have been issued for the evaluation of the working environment, which would strictly establish the procedures for their assessment, or determine the degree, level, and conditions of their harmful effects on the ergonomic system. In general, the following methodology can be followed:

- Elaboration of risk assessment, categorization, the decision on risky works; time frame of exposure and description of work activity; the processing of measurement protocols and comprehensive evaluation of factors of the working environment.
- Risk analysis carried out on the basis of a questionnaire.
- Detailed analysis of reported occupational diseases and risks of occupational diseases.
- Analysis of the health status of employees performed on the basis of data provided in medical reports on health fitness for work.

To eliminate harmful factors of the working environment, or to reduce them to the lowest possible level, it is necessary to take a whole range of technological, technical, organizational, and other measures. Despite the adoption of these measures, physical, chemical, biological, and other factors occur in the workplaces, which need to be analysed, assessed, evaluated, and subsequently measures taken to eliminate them. One of these factors of the working environment is lighting, which is the subject of this publication.

2 Ergonomics and Lighting in Workplace

Already in 1988, statistics from the World Health Organization reported that a person spends on average up to 90% of their life indoors. This type of environment requires particular attention in order to improve well-being, which includes visual well-being. It follows that a lighting device is a means which largely affects the nature of the environment in which it is installed. Its glow in the surrounding environment triggers a number of physiological and psychological reactions in man. That is why the term light microclimate is often used in this connection, which is created precisely by how a light device affects the environment that surrounds man. The basic task in lighting technology is to create a suitable environment with the best possible light microclimate, which helps to maintain healthy environment not only in the workplace, but also in other work and non-work areas. This chapter deals with the basic attributes of lighting within optics, with detailed focus on the typology of light and lighting of work environments in terms of ergonomics and determination of criteria for assessment within internationally applicable standards.

2.1 LIGHT AND LIGHTING

Light is physically characterized as an electromagnetic wave with a wavelength ranging between 380 and 760 nm, causing a physiological perception in the visual organ. More important for description of waves of light is the electric field vector. The magnetic field vector is not so important, because the induced magnetic field in the material can be neglected, as long as the magnetic movement of the electrons and the core is very slow for following

DOI: 10.1201/9781003359937-2

rapid, optical oscillation. It is an important factor in creating light well-being in the workplace.

Natural light, that is, light emitted, for example, from a flame or the sun, is unpolarized. It is a superposition of a large number of wave quanta, with each wave having a different frequency, amplitude, and direction. This results in a wave formation the vector of which is of a random size and – with the same probability – of any direction perpendicular to direction of the wave propagation. If the oscillation direction of the electric or magnetic vector or its magnitude is limited, a polarized light is produced. The polarization of light is characteristic of electromagnetic waves. Waves polarization occurs most commonly by reflection, refraction, or passing of light through an anisotropic environment. Linear polarization, circular polarization, and elliptic polarization are considered to be the basic types of polarization.

Typical of linearly polarized undulation is a zero-phase shift, that is $\varphi = \pm m\pi$ ($m = 0, \pm 1, \pm 2, \ldots$), which means $\cos\varphi = \pm 1$ a $\sin\varphi = 0$, that is $\frac{\varepsilon_{1x}}{E_{10}} \pm \frac{\varepsilon_{2y}}{E_{20}} = 0$ for two straight lines. Circular polarization occurs when constant amplitudes become equal ($E_{10} = E_{20} = E_0$) thus producing $\cos\varphi = 0$ and phase shift is $\varphi = \frac{\pm\pi}{2} + m\pi$ ($m = 0, \pm 1, \pm 2, \ldots$), on the basis of which the equation for circular polarization is expressed as follows:

$$\varepsilon_{1x}^2 + \varepsilon_{2y}^2 = E_0^2 \tag{2.1}$$

If the amplitudes are different and, at the same time, $\varphi = \pm m\pi$ ($m = 0, \pm 1, \pm 2, \ldots$), then the light is polarized elliptically. The light is also polarized elliptically if the amplitudes are equal, however $\cos\varphi \neq 0$. The equation for the elliptic polarization of light is identical to the ellipse equation and mathematically it is expressed as follows:

$$\left(\frac{\varepsilon_{1x}}{E_{10}}\right)^2 + \left(\frac{\varepsilon_{2y}}{E_{20}}\right)^2 - 2\left(\frac{\varepsilon_{1x}}{E_{10}}\right)\left(\frac{\varepsilon_{2y}}{E_{20}}\right)\cos\varphi = sin^2\varphi \tag{2.2}$$

In terms of ergonomics, light performs the following basic functions:

- visual function resulting from the physiology of human sight and the means of processing visual stimuli; visual functions – processing visual stimuli, perception: visual performance, how visual tasks affect visual performance, cognition, motor skills,
- system of perception – such as perception of visual discomfort and general perception of the environment and reception of information from there,

- non-visual functions – necessary to ensure biological processes in the body – circadian system.

2.2 CLASSIFICATION AND TYPOLOGY OF LIGHTING AT THE WORKPLACE

Due to characteristics of the space, the lighting falls into two basic categories:

- *interior lighting* – from the industrial point of view, it includes lighting of internal parts (office premises, industrial halls, warehouses, etc.). When designing it, the light reflected from all surfaces (walls, ceiling, floor, work surfaces) needs to be considered. One of the methods for computing this type of lighting is the so-called zonal cavity method, based on numerical determination of three-room spaces – the space between the ceiling and the furnishing, the space between the work plane and the floor, and the space between the furnishing and the work plane. Following individual calculations, the effective reflectance and utilization factor are determined from the tables, and ultimately determined is the average illumination value.
- *outdoor lighting* – from the industrial point of view, it includes lighting of adjacent parts of industrial premises, parking lots, truck loading areas, etc. It is used for illuminating space between individual buildings and support facilities. It must be designed in accordance with standards and, at the same time, it must not lead to a needless waste of power. In addition, it must not be used in places where it could be disruptive. When designing outdoor lighting, care must be taken to ensure adequate brightness, suitable shading, and an adequate amount of lighting suitably spaced apart.

On the basis of the origin of light, it is possible to determine the classification for artificial lighting, daylights, and combined lighting.

2.2.1 Artificial Lighting

Artificial lighting is used to create a light climate at a time when daylight cannot be used or is insufficient. A characteristic feature of artificial light is its relative stability over time. The primary advantage may include the possibility of its modification and use in the given space as needed. The disadvantage is its different spectral composition from daylight and thus its influence on the perception of colours. Ensuring the optimal level of illumination depending on the intensity of work increases the speed of resolution, which, for example, also reduces the time of execution of the task. Work performance improves, visual fatigue decreases, and, last but not least, injuries at work are prevented.

In terms of the power supply, artificial lighting can be categorized as follows:

- Normal lighting – main power supply
    - main lighting – used for work under normal operating conditions
    - auxiliary lighting – for auxiliary work outside the main operation
    - safety lighting – used in the event of technological equipment failures
- Failure lighting – powered from a backup source
    - backup lighting – allows work to continue in the event of malfunctions of the main lighting
    - emergency lighting – allows people to leave the area safely in the event of the main lighting failure

With regard to concentration of light:

- Total illumination – means lighting of a certain space without regard to specific requirements in its individual parts
- Scaled illumination – average intensity of lighting in individual parts of a given space depends on the visual intensity of the activity performed
- Local illumination – complements the total lighting
- Combined illumination – combination of the total and local lighting

Artificial lighting can also be categorized in terms of the luminous flux distribution of the lights used as follows:

- direct illumination – sharp, hard shadows are created, application is possible as auxiliary lighting to illuminate the work area
- mostly direct illumination – shadows with sharp contours are created, it is effective and applicable to areas with more visually demanding work

- mixed artificial lighting – soft and clearly defined shadows are created with no glare subject to correct direction of the lamp; it can be implemented in a work environment with more visually demanding work
- mostly indirect artificial lighting – a small share of shielding is created, but the emergent shadows are soft, this type is suitable for providing lighting of office areas
- indirect artificial lighting – luminous flux of the light fixture is directed upwards, no shielding is created, this type is not suitable for lighting of the work environment

The quantitative characteristic of artificial light sources is their specific power considered to be one of the primary operating parameters and indicating the proportion of the light output in relation to the power input of the light source. Increasing value means more light at lower power input, i.e., lower heat losses. The mathematical expression is as follows:

$$M_z = \frac{\phi}{P} \tag{2.3}$$

where:

M_z – specific power [lm/W]
φ - luminous flux of the source [lm]
P – input power of the power source [W]

Based on the main quantitative characteristic – the specific power – it is also possible to determine the efficiency of artificial light sources, according to the following mathematical equation:

$$\eta = \frac{M_z}{K_m} = \frac{\phi}{P \cdot K_m} \tag{2.4}$$

where

K_m – maximum light-technical efficiency of monochromatic radiation [lm/W]
η – efficiency of converting electric into light power [%]

Colour temperature, which describes the chromaticity (the colour perception) of the light emitted by an artificial light source, is also considered an input parameter of the artificial light source. This parameter is compared with the colour temperature of an ideal black-body radiator heated to this temperature. Incandescent light sources use the equivalent colour temperature "T_e" and discharge and LED light sources use the substitute colour temperature "T_n".

The substitute colour temperature is the corresponding point located at the nearest possible point of the temperature source showing chromaticity.

Perception of colours of objects is subject to spectral composition of the light source, the spectral reflection or transmittance factor of the object and the chromatic adaptation of the eye. Each eye has a different colour sensitivity, which is thus perceived differently. Therefore, the term "Colour Rendering Index" (CRI) has been introduced to detect the influence of each light source. The following mathematical expression may be used to determine the general colour rendering index:

$$R_a = \frac{1}{8}\sum_{i=1}^{8} R_i = \frac{1}{8}\sum_{i=1}^{8} (100 - 4.6\Delta E_i) \tag{2.5}$$

where:

R_a – general colour rendering index
R_i – special colour rendering index
ΔE_i – resultant colour shift

The maximum colour rendering index is 100, representing the same colour coordinates for 100 colour samples illuminated by the reference and the tested light source.

In order to express the specific power of the light source, it is also necessary to determine the photometric quantity of the luminous flux, which expresses the ability of the luminous flux to produce a luminous perception. The radiant flux φ_e is the power transmitted by radiation, i.e., the energy transmitted by radiation per unit of time. Mathematically, the luminous flux is expressed as follows:

$$\phi = K_m \cdot \int_0^{\infty} He(\lambda) \cdot V(\lambda) d\lambda \tag{2.6}$$

where:

$He(\lambda)$ – spectral density of the radiant flux at the point λ
$V(\lambda)$ – the relative luminous efficacy of monochromatic radiation, mathematically expressed as follows:

$$V(\lambda) = \frac{K_\lambda}{K_m} \tag{2.7}$$

where:

K_m – maximum light-technical efficiency of monochromatic radiation
K_λ – luminous efficacy of monochromatic radiation

2.2.2 Day Lighting

In order to ensure necessary day lighting, basic lighting systems are created in the design stage – the side lighting system, the upper lighting system, and the combined lighting system. The side system of day lighting consists of openings in the building – windows, which are located in the outer walls. Application of the side system of day lighting is not very effective of itself, as only parts of the space near the building openings are illuminated at the required values. The side system of day lighting may be divided as follows:

- unilateral side day lighting system – can be applied to a work environment with a higher permissible degree of unevenness of lighting, while protection against glare is ensured through northern orientation of the windows
- bilateral side day lighting system – can be applied to a work environment where permanent lighting conditions are required

The upper system of day lighting is provided by upper lighting openings – skylights, through which it is possible to achieve average day lighting without increased fluctuations. The ceilings of industrial halls with extensive workspaces are a frequent application.

The combined day lighting system is created by combining the side and the upper lighting system. It is most often applied in single-floor spaces, with the upper system reducing the unevenness caused by the side system.

2.2.3 Combined Lighting

In combined lighting, natural daylight is complemented by artificial lighting. Its purpose is, inter alia, to reduce the lack of daylight caused by the variability of external atmospheric conditions. The necessary level of illumination directly at the site of the visual task is facilitated by local illumination. It allows for customization and control of the illumination intensity according to the visual perception need.

Combined lighting is implemented in particular:

- in industrial halls and buildings where this type of lighting is required by technological conditions with clearly demonstrated lack of daylight,
- in industrial establishments which do not require the workers to be present permanently, which is about 50% of working hours,

- in interiors such as changing rooms, chambers, halls, shops, canteens, restaurants, cafes, etc.,
- in residential buildings in rooms such as a kitchen, hall, bathroom, toilet, etc.

Based on duration of use, combined lighting can be divided as follows:

- Permanent – artificial lighting is used in the interior of the building throughout the day. It is applied where people are permanently present but for serious reasons, whether functional or economic, it is not possible to obtain satisfactory day lighting.
- Transient – artificial light is used in the interiors with otherwise satisfactory daylight only for a certain period of time. This time depends mainly on the change in external lighting, which is caused by, for example, atmospheric conditions, dawn, dusk, etc.

According to the range, the combined lighting can be sorted into the following types:

- Total combined illumination – auxiliary lighting of all or a substantial part of the interior area by artificial radiance.
- Locally clustered illumination – in selected indoor areas with satisfactory day lighting, where access to daylight is restricted or where increased illumination is required for certain activities. An example may be rooms where technical devices cast a shadow on the work surface, or the shape of objects is differentiated (inside of hollow objects, etc.).

2.3 COMPLEXITY OF LIGHTING DESIGN AT THE WORKPLACE

The latest edition of Lighting of workplaces BS EN 12464-1:2021 Light and lighting supports complex solutions. When designing the lighting system of a workplace, priority should be given to creating optimal conditions for visual performance in a particular work environment so as to ensure maximum visual well-being. In view of the new possibilities of lighting technology, the standard has made some requirements stricter but, at the same time, the prescribed values do not apply to the entire space of the interior. The required lighting intensity, maximum permissible glare, or optimal colour rendering under particular light sources must be observed for individual types of work environments, especially at the points of the visual task and their immediate vicinity.

In production halls, maximum productivity is a priority and lighting is an important factor influencing it. Correct light distribution and light colour prevent unwanted glare and signs of fatigue among employees, also reduce the risk of accidents, facilitate visual tasks, and improve production efficiency. Gentle maintenance of light sources guarantees trouble-free operation and reduces the likelihood of production interruption.

In the automotive industry, most workplaces are located along assembly lines. Proper light distribution and light colour are critical factors for safe work and reliable production. The lighting requirements vary depending on the work tasks in the production process. Continuous linear illumination, arranged in parallel, is the best way to achieve sufficient illumination of the workplace at hand. Proper light distribution prevents distracting reflections and glare from shiny surfaces and makes it easier to perform tasks. Additional lights may be installed at the workplace to illuminate oblique work surfaces of the production line. It is important to remember that different tasks require different levels of illumination. It is, therefore, important to choose the intensity appropriately. Choosing the right lighting can significantly reduce power consumption, especially with light sources that are on all day.

Industrial enterprises are often open 24 hours a day, so well-designed, energy-efficient lighting that is uniform throughout the space plays an important role. Aisles may use presence detection technology to illuminate only the handling area where illumination is needed – the ability of the LED bulb to illuminate quickly ensures safety. The light source of the future is discussed at present, which is slowly but surely starting to occupy an exclusive position in all areas of lighting, whether it is interior lighting systems, exterior lighting, whether it is passive decorative lighting of the space, or a use in 100% functional lamps meeting all the requirements for proper spatial lighting. New materials and technologies are being developed, LED quality is improving and production costs – and thus the final price – are being cut. The LED is based on a semiconductor chip connected to a power source. It is surrounded by a layer of material (e.g., resin) which gives the emitted light the necessary optical properties (spot or scattered light, with different angles of illumination). As follows from the principle of a semiconductor diode, unlike a light bulb, where the direction of the passage of electric current does not matter, the LED must be connected only in the permeable direction. Thus, while the bulb can use both direct and alternating current, the LED can only use direct current. To connect it to an AC power source (for example to a normal socket), the LED lamp must be equipped with a rectifier. The LED emits light in a relatively narrow part of the spectrum – the monochromatic. The colour of the LED-emitted light, i.e., its wavelength, depends on the chemical composition of the semiconductor. LEDs are produced with different wavelengths ranging from ultraviolet, through different colours of the visible spectrum, to infrared. It took quite a long time to develop the blue LED. However, as soon as LEDs were available in the entire colour spectrum, a wide range of opportunities for practical use, including modern large-area colour screens, opened up to this light source. It was also not easy to develop white LEDs, which today are mainly used as a source of light in various light fixtures, lamps, and reflectors, as well as for backlighting liquid crystal displays.

Open spaces can benefit from intelligent control systems, which means that areas are illuminated only when necessary. The system detects presence in the aisle and then turns on the lights. After the set time has elapsed, if the presence is not sensed, the lights simply turn off. Reducing energy costs and maintenance time is mainly addressed in hard-to-reach and high-ceiling areas. Devices requiring high ceilings are often complicated and thus resolving long service life with good performance and lighting coverage allows for reduction in the number of necessary light sources. Long service life saves costs and reduces interruptions necessitated by replacement. Proper lighting control in large areas, such as storage areas, can significantly reduce operating costs and increase staff comfort by combining daylight and presence detection. If daylight is available, artificial lighting is dimmed.

When designing day lighting for the indoor work environment, the day lighting factor as the operational value is determined. This value is determined as a percentage and is calculated as the sum of the following components:

$$D = D_S + D_E + D_i = 100 \cdot \left(\frac{E_{is}}{E_{eH}} + \frac{E_{ie}}{E_{eH}} + \frac{E_{ir}}{E_{eH}} \right) \tag{2.8}$$

D_s – the sky component of the factor expressing the ratio of the part of the interior illumination E_{is}, caused by the sky brightness, and the actual horizontal illumination of an unshaded outer horizontal plane. The component expresses the illumination of a point within the assessed plane in the interior and the element of the area source to the current horizontal illumination of the exterior from the sky, which is evenly cloudy. The sky component may be determined using the following equation:

$$D_S = \frac{\int L_{dS} \cos \vartheta \frac{\cos \psi}{l^2} dS}{E_{eH}} \cdot 100 \tag{2.9}$$

ψ – angle of incidence of light rays on the illumination opening
ϑ – angle of incidence of rays on the spot under assessment
L_{ds} – luminance of the sky element with elevation angle determined at $\varepsilon = 90° - \vartheta$, calculated as follows:

$$L_{dS} = L_Z \frac{1 + 2\cos\vartheta}{3} \tau_0 0,\ 5\cos\psi(3 - cos^2\psi) \tag{2.10}$$

since the horizontal exterior lighting is a function of the zenith brightness Lz, it can be defined by using the following equation:

$$E_{eH} = \frac{7}{9}\pi L_Z \tag{2.11}$$

and thus, the general mathematical model for the sky component has the final form of:

$$D_S = \tau_0 \frac{100}{\pi} A_1 A_2 \int (B + \cos\vartheta)(C - cos^2\psi)\cos\vartheta \frac{cos^2\psi}{l^2} dS \tag{2.12}$$

where:

A_1, A_2, B, C – coefficients determined from the type of glazing and the brightness gradation of the sky

D_e – reflected external component of the factor expressing the ratio between the illuminated part E_{ie} (light reflected from the external reflective surfaces, the obstacles located in front of the building opening – a window, and the actual horizontal illumination. The mathematical formula is as follows:

$$D_e = \frac{D_{SF}}{10k_\varepsilon} \tag{2.13}$$

D_i – reflected internal component of the factor expressing the ratio between the interior component of the light in the interior resulting from multiple reflections of daylight from the surfaces located in the interior of E_{ir} and the actual horizontal illumination. The mathematical formula is as follows:

$$D_i = \frac{\chi\tau_0 S_0}{S(1 - \rho)}(D\rho_d + H_{ph}) \tag{2.14}$$

τ_0 – light loss factor
χ – correction factor
S – total interior surfaces area
S_0 – area of storage facility taken up by illumination opening
D, H – constants
ρ – average reflection factor of all internal surfaces
ρ_d (ρ_h) – average reflection factor of the lower (upper) part of the room, indicated above the horizontal plane, passing through the centre of the illumination opening.

In a combined lighting system, it is important to use lamps having the same or similar spectral composition as that of the daylight. Currently, white fluorescent lamps are the most suitable for this type of lighting. The requirements of combined lighting are as high as those of the previously mentioned lighting types. Therefore, both qualitative and quantitative requirements must be met.

In this type of lighting, the values of illumination by daylight and by artificial radiance are added up. It is, therefore, necessary to introduce common values and units. For the conversion of the day lighting factor under a cloudy sky, the value of the outdoor horizontal lighting is used as a different value in winter (5 000 lx) and in the summer (20 000 lx). As in the case of day lighting, the qualitative criteria for the light condition of the indoor environment must be met in both summer and winter, while the quantitative criteria must only be satisfied in winter. However, these requirements may result in disproportionately high demand for artificial lighting, especially in the summer months. It is therefore necessary to limit the brightness of the illumination openings with blinds, fixed screens, or special glazing. Additional illumination produced by the E_{di} artificial light in the case of side day lighting may be determined in different ways:

- From the condition of achieving the necessary level of lighting in winter months according to the equation:

$$E_{di} = E_p - 50e \tag{2.15}$$

  where
  E_p – is the illumination required for the activity at hand under artificial lighting,
  e – is the value of the day illumination factor achieved on that spot
- From the condition of maintaining uniformity in the summer months, corresponding to the requirement for side lighting by daylight, mathematically expressed as:

$$E_{d2} = 200(e_{max}r - e_{min}) \quad (2.16)$$

where
e_{min} (e_{max}) – is the smallest (largest) value of the day lighting factor achieved in the work plane in a room or its functionally defined part
r – the required uniformity of illumination

- From the condition of maintaining adequate brightness contrast in the field of vision, where crucial is the brightness of the sky as perceived from the interior through the window opening. Shading of the window with external obstacles is beneficial here. The mathematical expression is as follows:

$$E_{d3} = 500e_p(1 - 0.8\sin Z) \quad (2.17)$$

where:
e_p – value of the day lighting factor required by the standard
Z – average angle of interior shading from the central point of the window opening

3 Digitization and Digital Transformation in the Field of Ergonomics

The rapid development of technologies, as well as many changes in the global market, has led to a new trend – digital transformation. In a general context, digital transformation is a comprehensive transformation of any organization so that work in it is optimally carried out by information systems and software tools that together form a unified, fully integrated environment. Digital transformation can currently be considered a high-frequency activity for every area, affecting data analysis, design, process rationalization, etc. Digital transformation can currently be considered a high-frequency activity for every area, affecting data analysis, design, process rationalization, etc. In ergonomics, digital transformation has found its place precisely in assessing not only the physical workload of the workers but also in improving their working conditions, whether in designing administration offices or organizing production workplaces and a comprehensive assessment of the work environment. The chapter presents the specificities of the issue of digitization and digital transformation in its conceptual foundations with direct applicability in the ergonomic processes with a focus on the issue of lighting.

3.1 DIGITIZATION VS. DIGITALIZATION

Today, the term "digital" is synonymous with the speed of change that is taking place in today's world through rapid adoption of technology. In

DOI: 10.1201/9781003359937-3

information science, digitization can be basically understood on two fundamental levels. The first level is an expression of the specific physical transfer of analogue information to digital information, i.e., "digitization" – "transfer of information to digital (numerical) expression". The second meaning of digitization, i.e., the introduction of new digital technologies into life, is called "digitalization". Thanks to this terminological differentiation, terminologically vague definitions do not occur and it is possible to employ this abstract term in specific contexts. The two definitions used can be broken down by their frequency of use into "digitalization", which more often coincides with the English term "digitization", and "digital transformation", which best coincides with the English term "digitalization".

With regard to globally recognized definitions, digitization in the first sense (digitization) is understood as "the creation of digital (bits and bytes) versions of analogue/tangible things such as paper documents, microfilm images, photographs, sounds and others. This means converting or representing something non-digital into a digital format that can then be used by a computing system". According to them, digitized documents are subsequently used for secondary purposes in company processes, so digitization in this sense can also be understood as a certain automation of the existing manual processes. According to this source, digitization in the second sense is "the use of digital technologies and data (digitized and originally digital) to generate profit, improve business, replace/transform business processes (not just digitising them) and create an environment for digital commerce, with digital information being the basis".

The following are considered to be the main benefits of digitization:

- Online access to business opportunities and customers.
- The use of modern technologies and innovative services.
- Reduction in operation overheads.
- Simplifying business processes and increasing business efficiency.
- Expanding your company's operations with new business models.
- Simplified business abroad due to changes in legislation.
- Promoting consumer protection as well as protection from difficult customers.
- Competitiveness and modern way of delivering services to customers.

The basic differences between the digitalization and digitization are possible to present via short definitions with examples:

- *Digitalization*
  - Definition: Transforming business processes by leveraging digital technologies, ultimately resulting in opportunities for efficiencies and increased revenue

    - Examples: Analysing data collected by Internet-connected devices to find new revenue streams; using digital technology to transform your reporting processes, collecting and analysing data in real time and using insights to mitigate risk and promote efficiency on future projects
- Digitization
    - Definition: Converting data, documents and processes from analogue to digital
    - Examples: Scanning a photograph to create a digital file; converting a paper report to a digital file such as a PDF; inputting an existing paper checklist into a digital checklist app; recording a presentation or phone call, turning physical sound into a digital file

3.2 SPECIFICATION OF DIGITAL TRANSFORMATION

The term "transformation" means that the use of things digital integrally enables new types of innovation and creativity in a particular area. Digital transformation is defined as digital solutions of enterprise transformation, consisting of the following areas (Figure 3.1):

- Increasing the efficiency of the utilities used, reducing the cost of energy in the enterprise, and reducing emissions.
- Building smart cities with the help of technology and providing citizens with a higher quality of life.
- Protection of information and resistance to attacks on the communication network or corporate systems.
- Streamlining employee cooperation, simplifying administration, and accelerating processes in the enterprise.
- Increasing flexibility for employees and customers while increasing the effectiveness of interaction within the organization.
- Modernization of the enterprise through process automation, data use, and production processes efficiency.
- Freedoms of the decision-making of employees lying in where, how, and when to enhance their qualifications or receive mandatory training.
- Getting rid of corporate IT as a non-productive part of the business and focusing on the business.

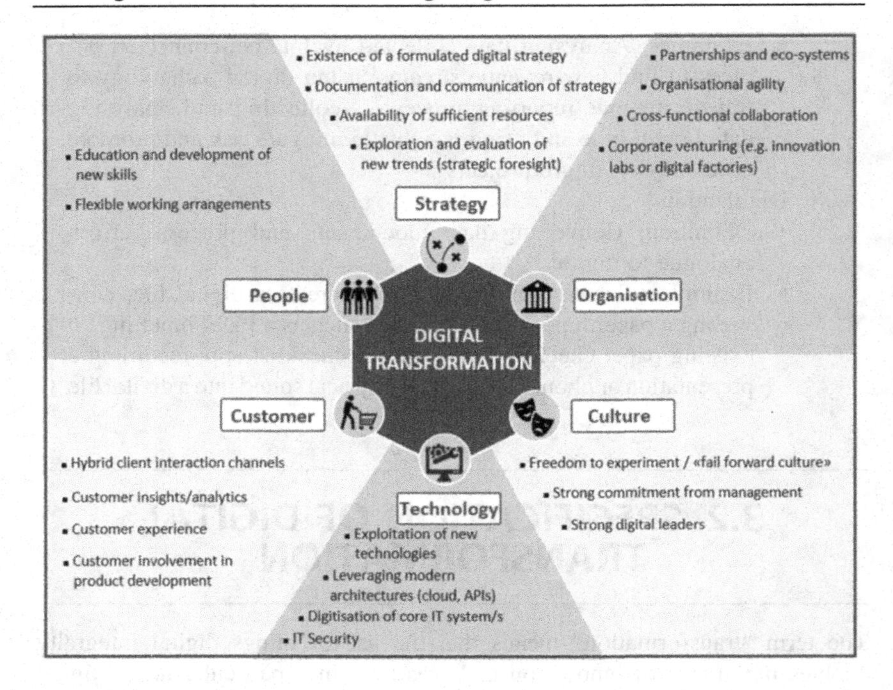

FIGURE 3.1 Conceptual framework of digital transformation.

In a general context, digital transformation is a comprehensive transformation of any organization so that work in it is optimally carried out by information systems and software tools that together form a unified, fully integrated environment. Advantages of digital transformation include:

- Time saving – automation speeds up all processes and increases work performance.
- Reduction of time needed to introduce new services – suitable for companies that frequently change their portfolio; changing the service parameters requires only a rewriting in the tables.
- Fast response to customer requirements, flexibility – only a few minutes pass between sending a holiday destination offer to confirming an order, along with assigning the travel insurance number.
- Data-driven decision-making is made possible by the ability of the system to simply answer business questions.

For the appropriate characterization and specification of terms, a general summary of the differences between the terms is given in Table 3.1.

TABLE 3.1 Summary of differences within the issue

	DIGITALIZATION	*DIGITIZATION*	*DIGITAL TRANSFORMATION*
Focus	Processing of data	Transformation of data	Taking advantage of information
Goal	Automate business processes	Convert analogue progress to digital	Changing the way companies operate and think
Tools	IT systems and applications	Computers and other equipment for converting	New digital technologies
Activity	Creating a fully digital workflow process	Digitizing physical resources such as paper documents, photos, videos	Creating a new digital company
Challenge	Financial	Material	Human resources

3.3 DIGITALIZATION PROCESSES IN ERGONOMICS

Using digitalization of production and 3D designing is the most common form in the compilation of the so-called Digital twin, which is also part of the concept and the idea of Industry 4.0. The digital twin can be understood as a digital sensor model of a real object within the simulation process at active deployment. The basis for the digital twin is cumulative massive data collected in real time within the real environment. The collected data form the basis for the creation of objects and processes within the digital environment, which in turn provides information on performance, guiding subsequent changes in real processes and systems. The digital twin concept has undergone a gradual formation that has been integrated into the four basic stages. The first stage is focused on "information mirroring model" – conception of digital twin during the 18 years (1985–2002). The second stage is "Digital Simulation, 3D printing" is focused on collaboration, simulation, and workflow across global enterprises during 12 years (2003–2014). The third stage "intelligent & connected IoT services" provides connection of devices with using big data analytics during three years (2014–2016). The last stage "mixed reality, cognitive and AI" is focused on artificial intelligence, cognitive services, and holographic from 2017 to present.

The above stages are interconnected and used depending on the required attributes of the issue. The following tasks are required for proper implementation of the digital twin:

- objects creation (2D and 3D) – production systems, products, halls,
- collection of real data from the analysed operation,
- implementation of simulation tools with continuous evaluation of the obtained data,
- verification of the results achieved with subsequent implementation in a real environment.

If the digital twin model is constructed correctly, the following advantages can be achieved within the production concept, which this concept caters to:

- continuous collection and evaluation of information and data,
- elimination of costs arising from error and waste,
- makes managing the course of the process from production to logistics to optimization easier,
- ensuring that relevant information is provided within a short period of time.

The issue of digitalization lighting of work environment is based on standard thermodynamic, optical, or light-technical lighting models. The Monte Carlo method and radiation method (Figure 3.2) are considered to be the basic methods that are implemented in computer programs focused on analysis, design, and assessment of lighting of the work environment. The Monte Carlo method applies the principle of ray tracing. The physical nature of both methods is similar, but the difference occurs in the algorithmizing and size of the surfaces, because in the Monte Carlo method of ray tracing, work is carried out stochastically based on micro surfaces, while the radiation method works deterministically at the larger surfaces level. Computer programs that use the principle of ray tracing are capable of accurate modelling of lighting elements.

The Monte Carlo method principle is applied mainly to digital models, featuring surfaces with diverse optical properties. The reason for the implementation of this method in computer visualizations of lighting is its general applicability with the possibility of controlling the accuracy and time of machine calculation. The method uses a number of particles or rays that move in the environment and are simultaneously subject to basic physical laws. The concept of particle tracing is used to track the light propagation from the source to the environment or to the observer, but this is considered to be disadvantageous. The reason for the disadvantage of this type of tracking is the loss of a large number of

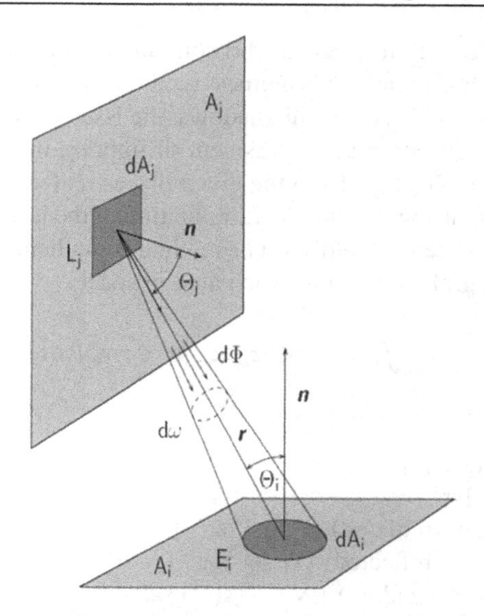

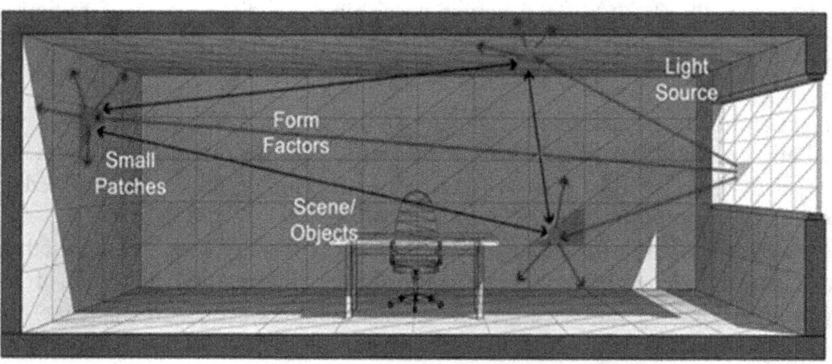

FIGURE 3.2 Illustration of ray tracing (up) and principal representation of the radiation method (down).

rays or particles that travel through space to the observer. In view of this fact, the "back ray tracing" method is more often used, which works in the exact opposite way, i.e., from the observer to the source.

The calculation algorithms of the Monte Carlo method are proportionally dependent on the amount of light particles and simultaneously on the luminous flux density carried by each of these particles. If the number of particles emitted is large, the method of their tracing is not important, so it is possible to apply algorithms based on standard particle tracing or back tracing.

For places that are illuminated in the computer model directly, their so-called direct illumination or default brightness is calculated. In places illuminated partially, it is possible to observe semi-shadows, the issue of which is addressed in simulation tools focused on the assessment of lighting in various ways. An example is the emission of rays from the given places under a principle of randomness towards a light source with the illumination of the intersections created proportional to the degree of shielding. The method is mathematically described by the following integral equation for each area in space:

$$L_r(\theta_r, \Phi_r) = L_e(\theta_r, \Phi_r) + \iint L_i(\theta_i, \Phi_i)\rho_{bd}(\theta_i, \Phi_i, \theta_r, \Phi_r)|\cos\theta_i|\sin\theta_i d\theta_i d\Phi_i \quad (3.1)$$

where:

Φ – azimuth angle [°]
θ – polar angle [sr]
$L_e(\theta_r, \Phi_r)$ – glow (radiated) [$W.m^{-2}.sr^{-1}$]
$L_r(\theta_r, \Phi_r)$ – glow (reflected) [$W.m^{-2}.sr^{-1}$]
$L_i(\theta_i, \Phi_i)$ – glow (incident) [$W.m^{-2}.sr^{-1}$]
$\rho_{bd}(\theta_i, \Phi_i, \theta_r, \Phi_r)$ – a function describing the distribution of radiation transmission and reflection dependent on direction of the reflected and the incident radiation [sr^{-1}].

If generalized, the above method of ray tracing may be understood as the radiation method in a certain way. Principally, the radiation method considers the surfaces to be ideal diffuse light radiators. This principle was applied as one of the first in the field of light computing.

In terms of algorithmizing and visualization, the advantage of the method consists of independent calculation of surface illumination in terms of direction of viewing an object, a space, or a complex scene. In this case, the mathematical expression is realized by means of the radiation equation of an area element i as follows:

$$M_i = M_{0i} + \rho_i \sum M_j A_j F_{ji} / A_j \quad (3.2)$$

where:

M_{0i} – own emission of the element i
ρ_i – element light reflection factor i [-]
M_j – description of light emanating from the element j
F_{ji} – element shape factor from j to i [-]
A_j – element area [m^2]

Since the variable – the shape factor is considered to be a dimensionless geometric value, it depends solely on the relative position of individual surfaces in space and the shape of the given surfaces. This variable is also a relative quantity expressing the amount of reduction of diffuse radiation of the element j incident on the element i. Based on this fact, equation (3.2) can be mathematically simplified into the following form:

$$M_i = M_{0i} + \rho_i \sum M_j F_{ji} \tag{3.3}$$

The solution of the system are the obtained necessary values of the emitting lighting for each area element. Simulation tools that are currently used digitalising the lighting of workspaces most often use a combination of both of the two methods described above. Some simulation tools also provide a calculation method option, which, however, requires experience and great expertise in a given issue on the part of the user, as there may be a risk of errors made in the simulation due to inappropriate selection of the computing simulation method. In 2020, a study published in the journal *Solar Energy* was carried out, focused on the overview of light transport algorithms used in simulation tools in interiors. It was presented a graphical interpretation of the light-technical instruments that have been on the market since 1970, indicating their computing method – radiation, back ray tracing, leading beam tracing, and bi-directional algorithm.

As part of the digital transformation, it is possible to mention the Dialux Evo software solution as the main protagonist, developed by the German company Dial GmbH and adapted to design, visualize, and computation of lighting for individual rooms, floors, buildings, and exteriors. To create artificial lighting, it uses catalogues of light fixtures from various manufacturers, which are either available online on specific websites or installed by the user as active plug-ins. The individual light fixtures catalogues include classic, incandescent, fluorescent, street, halide, LED lighting, optical fibres, optics, and special projectors. In order to quickly create the model to be evaluated, this tool enables the import of a layout solution from CAD software or it is possible to draw the whole simulation model directly in the program and then convert it to a CAD file. This light-technical tool is used by more than 680,000 designers around the world to model and simulate artificial, daylight, and combined lighting.

Based on the selection of the modelled type of work environment, the corresponding software work environment is started. This environment consists of five basic areas used to create a project, including definition of elements for working with the model, a toolbar, a space for visualizing the project created a space for watching the simulation and the results achieved, and additional settings.

Defining the computation parameters is implemented by means of objects located on the computation objects panel. Once they are defined, the simulation itself starts, followed by the subsequent report analysis of the results achieved. The achieved results can be analysed in an interactive form via a software interface, or in the form of an electronic PDF/print output.

Lighting in a digital environment is analysed in a similar way in the ReluxDescope software, which covers analyses, simulations, and subsequent evaluations of lighting in interior facilities, exterior facilities, and assessment of roadway lighting. This tool ensures both CAD and BIM compatibility, supporting generally applicable legislation and standards (EN, DIR, ASR, etc.). The tool is currently actively used by more than 220,000 registered users, and in the basic module it functions as a freely available software solution with the possibility of purchasing modules for AutoCad, Swiss energy standards, or, for example, lighting computing module for tunnels, complete with the current legislation.

The opening window of the program offers a selection of several lighting project planning options divided into two basic sections – interior lighting planning, exterior lighting planning.

Based on the selection of the relevant section and a specific type of a facility, it is possible to define lighting simulation parameters, with the room parameters, degrees of reflectance, light fixtures, lighting factors, specification of parameters for defining the daylight factor, etc. provided. Indicators of the result part of the program provide a numerical and graphical interpretation of the completed simulation with the possibility of exporting the results to external files or directly printing them. In the right part of the work environment is a project structure tree, which provides a comprehensive overview of the results achieved with an interactive possibility of switching between the parameters under assessment.

4 Implementation of Digital Ergonomic Transformation in the Engineering Industry

The fourth chapter presents a practical solution for the implementation of digital transformation in the engineering industry. Implementation of digital transformation in ergonomics is interpreted for the engineering industry by means of an example from practice – assessment and redesign of lighting of the work environment of a production hall. The first part of the chapter describes the implementation of in-situ measurements at selected production sites with the identification of values that do not comply with the prescribed minimum requirements for lighting in the performance of work activities. The second part describes the possibility of digital transformation in practice, using the creation of a digital twin production hall. The adequacy of the digital transformation is supported by a comparison of the achieved results of in-situ measurement and the digital model. The conclusion of this chapter is a description of the ergonomic rationalization solution, which confirms the complex suitability of the model for real application in practice.

DOI: 10.1201/9781003359937-4

4.1 DIRECT IN-SITU ILLUMINATION MEASUREMENT

Three measuring points were placed in the production hall with dimensions of 26 000 × 17 000 × 8 500 mm, presenting workstations at production facilities (Figure 4.1). At each measuring point, nine measurements were made 0.85 m above the floor – three in the vertical position, three in the horizontal position, and three at the auxiliary desk. Distribution of the individual measuring points is shown in Figure 4.1.

The illumination measurement was carried out with a standard digital luxmeter with a resolution of 0.1 lx. In the first case, the measured values were those of intensity of the combined lighting.

The data obtained from measuring intensity of the combined lighting were processed and recorded in Table 4.1 with the subsequent graphical interpretation (Figure 4.2) and assessment of the values measured.

The second measurement of illumination in the production hall recorded and evaluated the intensity of day lighting, from which the day lighting factor was subsequently calculated. The data obtained are interpreted in Table 4.2 and Figure 4.3. The third measurement of illumination recorded and evaluated the intensity of artificial lighting. Obtained results are provided in Table 4.3 and Figure 4.4.

The individual measurements and their graphical interpretations make it clear that none of the measurement points has reached the required value of illumination intensity and, therefore, it is necessary to make changes to the workplace.

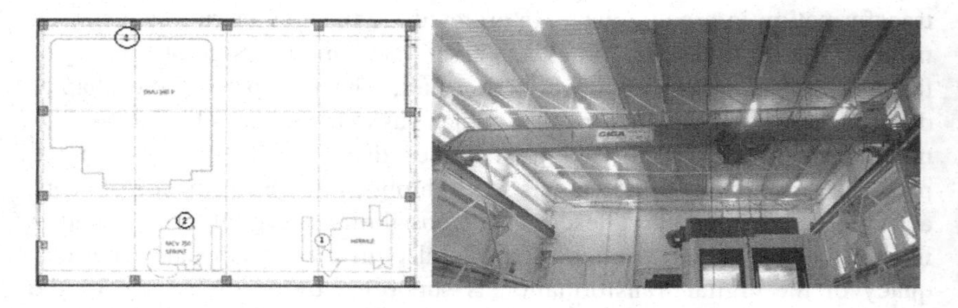

FIGURE 4.1 Distribution of measuring points (left) and production hall illumination under assessment (right).

TABLE 4.1 Results of direct measurement of the combined illumination intensity

MEASURING POINT	*1*			*2*			*3*		
NUMBER OF MEASUREMENT	*1*	*2*	*3*	*1*	*2*	*3*	*1*	*2*	*3*
E_{iz} [lx]	118.9	118.2	118.4	45.9	46.2	45.8	119.1	118.6	109.1
E_{iv} [lx]	74.8	64	73,5	128	130.4	116.9	167.3	169.0	166.9
E_{ip} [lx]	131.9	131.7	136.9	141.7	139.2	141.3	171.3	171.6	175.8
$\bar{E}_{iz}$ [lx]		118.5			46,0			115.6	
$\bar{E}_{iv}$ [lx]		70.8			125.1			167.7	
$\bar{E}_{ip}$ [lx]		133.5			140.7			172.9	
σ_{iz}		0.294			0.170			4.601	
σ_{iv}		4.814			5.880			0.910	
σ_{ip}		2.406			1.096			2.054	

Note:
E_{iz} – illumination intensity at a given point "i" measured in vertical position [lx],
E_{iv} – illumination intensity at a given point "i" measured in horizontal position [lx],
E_{ip} – illumination intensity at a given point "i" measured at the auxiliary desk [lx],
$\bar{E}_{iz}$ – comparative illumination for the exterior unshaded horizontal plane at a specific point "i" measured in vertical position [lx],
$\bar{E}_{iv}$ – comparative illumination for the exterior unshaded horizontal plane at a specific point "i" measured in horizontal position [lx],
$\bar{E}_{ip}$ – comparative illumination for the exterior unshaded horizontal plane at a specific point "i" measured at the work desk [lx],
σ_i – standard deviations from the values measured.

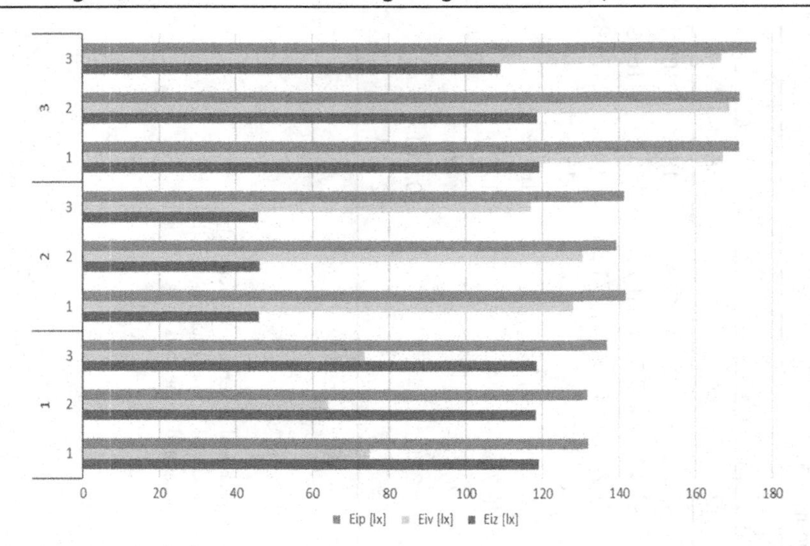

FIGURE 4.2 Graphical representation of direct measurement of the combined illumination intensity.

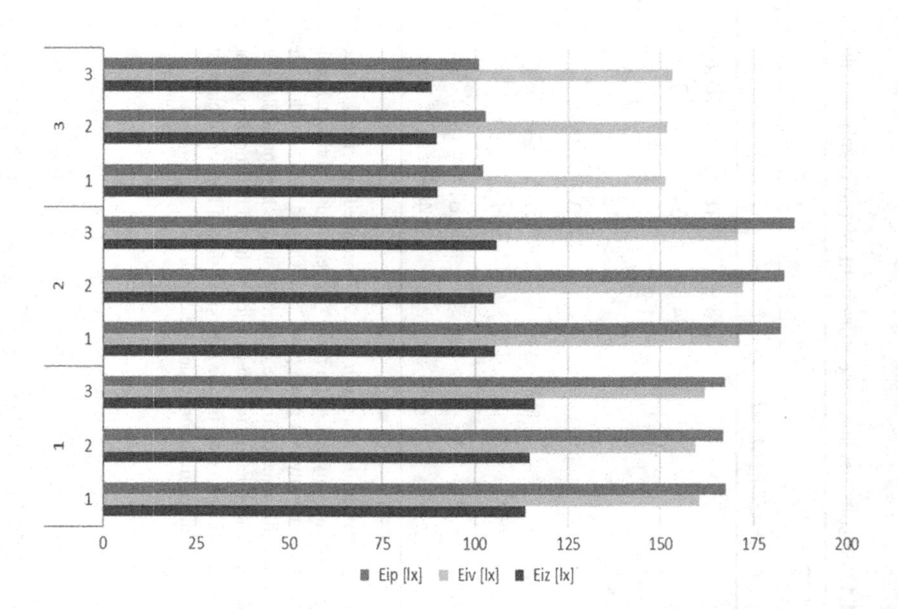

FIGURE 4.3 Graphical representation of direct measurement of daylight illumination intensity.

TABLE 4.2 Results of direct measurement of day lighting in the production hall

MEASURING POINT	*1*			*2*			*3*		
NUMBER OF MEASUREMENT	*1*	*2*	*3*	*1*	*2*	*3*	*1*	*2*	*3*
E_{iz} [lx]	12	13	12	4	6	6	12	9	5
E_{iv} [lx]	8	9	9	12	12	14	8	8	9
E_{ip} [lx]	5	4	4	9	11	10	5	4	4
E_{hz} [lx]	3347	3348	3351	3344	3341	3352	3352	3355	3351
E_{hv} [lx]	3353	3358	3352	3362	3361	3365	3362	3368	3360
E_{hp} [lx]	3355	3349	3351	3364	3367	3362	3357	3355	3359
σ_{iz}		0.471			0.943			2.867	
σ_{iv}		0.471			0.943			0.471	
σ_{ip}		0.471			0.816			0.471	
σ_{hz}		1.700			4.643			1.700	
σ_{hv}		2.625			1.700			3.399	
σ_{hp}		2.494			2.055			1.633	
D_z [%]	0.359	0.388	0.358	0.120	0.180	0.179	0.358	0.268	0.149
D_v [%]	0.239	0.268	0.268	0.357	0.357	0.416	0.238	0.238	0.268
D_p [%]	0.149	0.119	0.119	0.268	0.327	0.297	0.149	0.119	0.119

Note:

$\sigma_{i\ (h)}$ – standard deviations from the values measured,
D_z – day lighting factor for the vertical position [%],
D_v – day lighting factor for the horizontal position [%],
D_p – day lighting factor for the position at the work desk [%].

TABLE 4.3 Results of direct measurement of artificial lighting in the production hall

MEASUREMENT POINT	*1*			*2*			*3*		
NUMBER OF MEASUREMENT	*1*	*2*	*3*	*1*	*2*	*3*	*1*	*2*	*3*
E_{iz} [lx]	113.5	114.8	116.2	105.4	105.2	105.9	89.9	89.5	88.3
E_{iv} [lx]	160.4	159.3	161.7	171.1	172.1	170.7	151.3	151.7	153.2
E_{ip} [lx]	167.6	166.9	167.2	182.3	183.2	186.1	102.2	102.8	100.9
$\bar{E}_{iz}$ [lx]		114.833			105.500			89.233	
$\bar{E}_{iv}$ [lx]		160.467			171.300			152.067	
$\bar{E}_{ip}$ [lx]		167.233			183.867			101.967	
σ_{iz}		1.103			0.294			0.680	
σ_{iv}		0.981			0.589			0.818	
σ_{ip}		0.287			1.621			0.793	

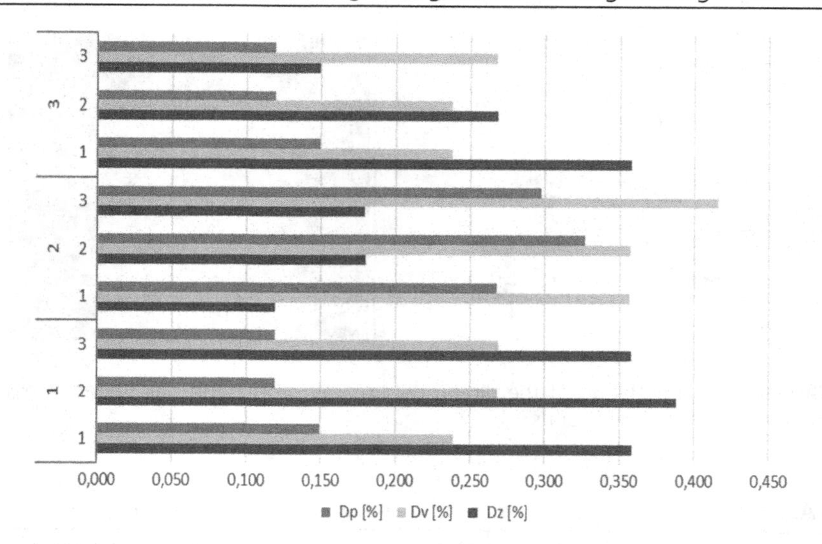

FIGURE 4.4 Graphical representation of direct measurement of artificial lighting intensity.

4.2 THE USE OF DIGITAL TRANSFORMATION TOOLS IN BUILDING A DIGITAL TWIN OF THE PRODUCTION HALL

The design and redesign of lighting for the work environment can be implemented through the tools of digital ergonomics. Based on the measurements of the production hall lighting, a digital twin model with three production machines (in accordance with the measured points and the hall layout) was compiled using the Diaux Evo simulation tool (Figure 4.5). The final comparison draws on the individual computation areas' average values.

Upon construction of the lighting system model of the production hall, a simulation of artificial lighting was carried out, the results of which are presented numerically in the Table 4.4 and graphically through the display of the course of isophotes (Figure 4.6).

Based on the results obtained, it .can be stated that both direct measurements of artificial lighting in situ and the digital twin simulation show on

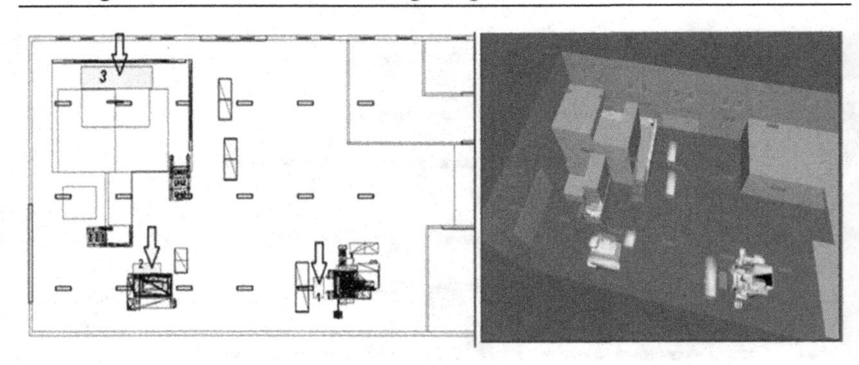

FIGURE 4.5 Rendering of the digital twin of the production hall under assessment, with measuring points.

TABLE 4.4 Results of the current state simulation – artificial lighting

CALCULATING AREA	*PARAMETER*	*AVERAGE VALUE [LX]*	*MIN. VALUE [LX]*	*MAX. VALUE [LX]*
Calculating area – Measurement point No. 1	Illumination intensity	133.25	67.5	199
Calculating area – Measurement point No. 2	Illumination intensity	167.5	133	202
Calculating area – Measurement point No. 3	Illumination intensity	120	93	147

average 65% lower values than the values required by legislation dealing with artificial lighting of production halls.

Day lighting model in Dialux Evo 6.1 compiled using an artificial lighting model with a change in basic simulation parameters. The simulation of day lighting (Figure 4.7) in the given program was compiled using a sky model – overcast sky. The programme determines the day lighting factor in four sky models. In this case, the sky model setting is only indicative, as the internal illumination intensity component has been considered in the assessment of day lighting (Table 4.5). This component was determined as a guidance, since it is not possible to set the value of the external illumination in the given programme and thus the resulting values of the day lighting factor

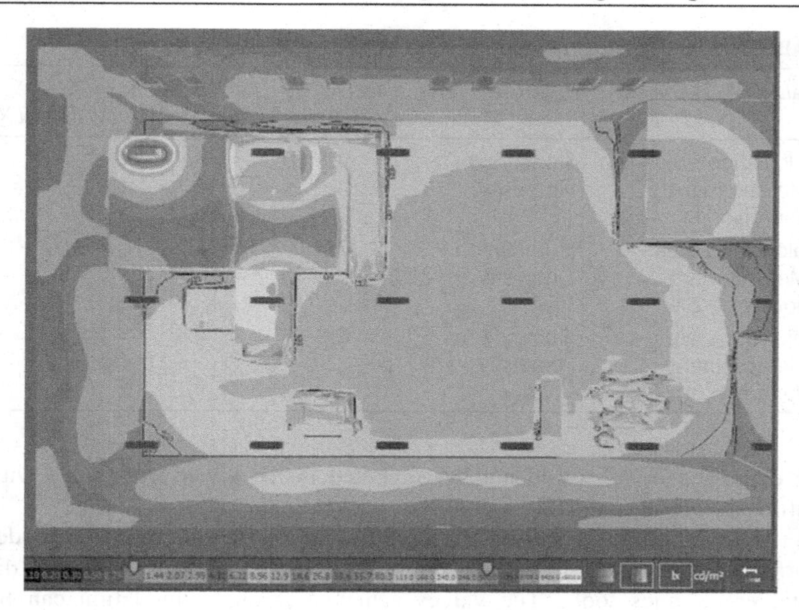

FIGURE 4.6 The course of the artificial lighting isophotes of the production hall's digital twin.

FIGURE 4.7 The course of isophotes of the day illumination of the production hall's digital twin.

TABLE 4.5 Results of the current state simulation – day lighting

CALCULATING AREA	*PARAMETER*	*AVERAGE VALUE [LX]*	*MIN. VALUE [LX]*	*MAX. VALUE [LX]*
Calculating area – Measurement point No. 1	Illumination intensity	9.1	8.4	9.8
Calculating area – Measurement point No. 2	Illumination intensity	7.3	4.2	10.4
Calculating area – Measurement point No. 3	Illumination intensity	6.8	4.4	9.2

D, expressed in %, would not be computed in the given programme with sufficient relevance.

By combining the settings of the artificial and day illumination model parameters, a model of combined lighting was created with the use of digital ergonomics tools. The values achieved through simulation can be found in Table 4.6 and the graphical course of the isophotes is presented in Figure 4.8.

In order to ensure the correctness of the assembled digital twin of the production hall and the need for further analysis with the subsequent rationalization of the complex lighting model of the production hall, the results of direct in-situ measurement and the results from the simulations were compared (Table 4.7) on the basic principle of average values.

TABLE 4.6 Results of the current state simulation – combined lighting

CALCULATING AREA	*PARAMETER*	*AVERAGE VALUE [LX]*	*MIN. VALUE [LX]*	*MAX. VALUE [LX]*
Calculating area – Measurement point No. 1	Illumination intensity	115	83	147
Calculating area – Measurement point No. 2	Illumination intensity	112.5	74	151
Calculating area – Measurement point No. 3	Illumination intensity	149.3	87.6	211

FIGURE 4.8 The course of isophotes of the combined illumination of the production hall's digital twin.

TABLE 4.7 Comparison of the results of in-situ measurements and the current state simulation

	ARTIFICIAL ILLUMINATION		
	MEASURING POINT NO. 1	*MEASURING POINT NO. 2*	*MEASURING POINT NO. 3*
Mean value of in-situ measurement [lx]	147.5	153.6	114.4
Simulation – Illumination intensity [lx]	133.25	167.5	120
Deviation – real measurement/ simulation [%]	**9.668**	**9.081**	**4.875**

	DAY LIGHTING		
Mean value of in-situ measurement [lx]	8.4	6.7	7.1
Simulation – Illumination intensity [lx]	9.1	7.3	6.8
Deviation – real measurement/ simulation [%]	**8.333**	**9.445**	**4.225**
	COMBINED ILLUMINATION		
Mean value of in-situ measurement [lx]	107.6	103.9	152.1
Simulation – Illumination intensity [lx]	115	112.5	149.3
Deviation – real measurement/ simulation [%]	**6.877**	**8.277**	**1.841**

4.3 THE USE OF DIGITAL TRANSFORMATION TOOLS IN THE ERGONOMIC LIGHTING RATIONALIZATION OF THE PRODUCTION HALL

Based on the results of the current state summarized in the previous chapter, it is necessary to implement ergonomic rationalization measures, which can be implemented by using the elements of digitalization. Due to practical requirements and financial possibilities, an alternative was created involving a comprehensive replacement of light fixtures, including their new layout with adjustment of their position in all axes, i.e., x, y, z.

The original light fixtures located in the facility under assessment – the production hall – were replaced with light fixtures whose technical specifications are given below:

- Input power: 50W
- Colour temperature: 4 000 K
- Luminous flux: 7 000 lm

- Ra: 92
- Operational efficiency: 99.98%

The basic rationalized model is identical to the digital twin in terms of its layout and dimension. As part of the digital elements verification, the model is analogously identical to the verification twin, that is, a correspondence of individual models between the settings of computing surfaces, measurement rasters, work planes, etc. exists.

By creating a rationalized model of the production hall's digital twin, it is possible to carry out an assessment of the basic qualitative and quantitative attributes of lighting, which include, for example, artificial lighting, intensity, aesthetics, or maintenance of the designed light fixture or building openings.

In the assessment of artificial lighting, the guiding factor is the parameter of the lighting intensity. This parameter is a quantitative value and is expressed in lux. The course of the illumination intensity isophotes is shown in Figure 4.9.

The table of simulation results of the artificial illumination intensity of the new proposal (Table 4.8) confirms the suitability of the proposed lighting in terms of the parameter studied, as at all computation points the illumination

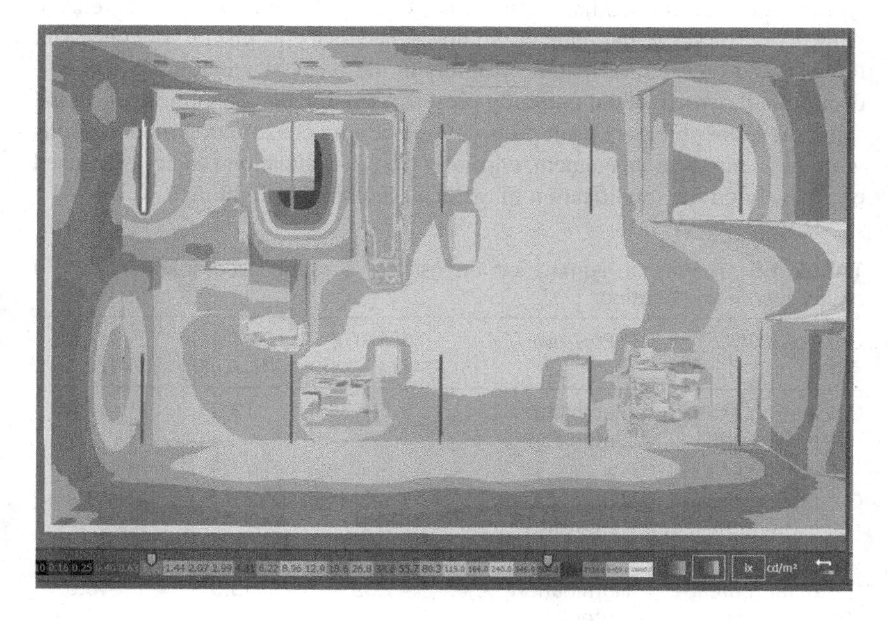

FIGURE 4.9 The course of the artificial lighting isophotes of the production hall's digital twin after rationalization.

TABLE 4.8 Results of illumination intensity of a rationalized solution using a digital twin – artificial lighting

CALCULATING AREA	*PARAMETER*	*AVERAGE VALUE [LX]*	*MIN. VALUE [LX]*	*MAX. VALUE [LX]*
Calculating area – Measurement point No. 1	Illumination intensity	333	312	354
Calculating area – Measurement point No. 2	Illumination intensity	335	303	367
Calculating area – Measurement point No. 3	Illumination intensity	355	326	384

intensity reached a higher than the legislative value and exceeded the framework of the standard.

Given that the production hall is in operation 24 hours each business day, the assessment of the intensity of day illumination (Table 4.9 and Figure 4.10) is purely informative, given that even on clear and sunny days it is not possible to work in such a designed hall with the use of only day illumination. To enable work using only the daylight, it would be necessary to install skylights, sun tunnels, and other construction openings in the form of windows in the production hall, which would currently present a significant economic cost. In the future, it is possible to consult proposals containing construction modifications with the company management, currently the rationalization project is focused exclusively on the modification of artificial (combined) lighting.

TABLE 4.9 Results of illumination intensity of a rationalized solution using a digital twin – day lighting

CALCULATING AREA	*PARAMETER*	*AVERAGE VALUE [LX]*	*MIN. VALUE [LX]*	*MAX. VALUE [LX]*
Calculating area – Measurement point No. 1	Illumination intensity	20.35	19.3	21.4
Calculating area – Measurement point No. 2	Illumination intensity	25.75	21.7	29.8
Calculating area – Measurement point No. 3	Illumination intensity	41.05	33.5	48.6

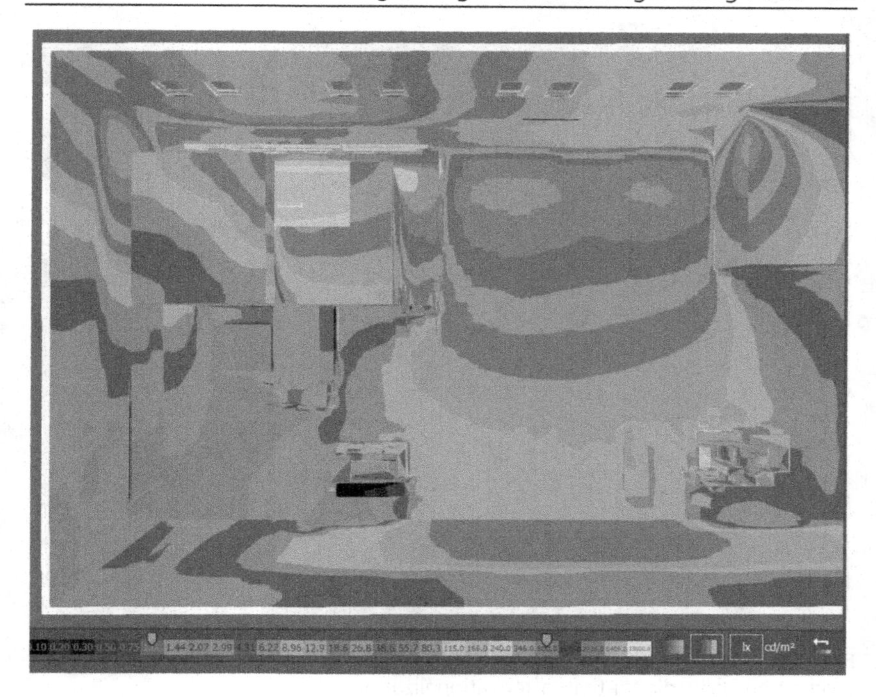

FIGURE 4.10 The course of isophotes of the day illumination of the production hall's digital twin after rationalization.

As was the case with artificial lighting, the primary parameter for the assessment is the intensity. Based on the results obtained (Figure 4.9–Figure 4.11, Table 4.8–4.10), it can be concluded that the average values from simulations of the rationalized digital twin model of the production hall fully satisfy the legislative and standard frameworks.

It is also possible to comprehensively assess the qualitative parameter – aesthetics, which is ensured by the installation of a new modern light fixture designed specifically for the industry. This new modern light fixture has a precisely defined maintenance factor, which was also determined in the light-technical computations. This factor's coefficient for the proposed source of artificial lighting is 0.8. To ensure maintenance of construction openings, a company manual has been developed, defining the exact maintenance intervals.

On the basis of a comparison of the in-situ measurements of the current state and the proposed rationalization solution (Table 4.11), it is possible to determine the suitability of the proposed ergonomic rationalization in the form of a reduction and simultaneous replacement of the old lighting system.

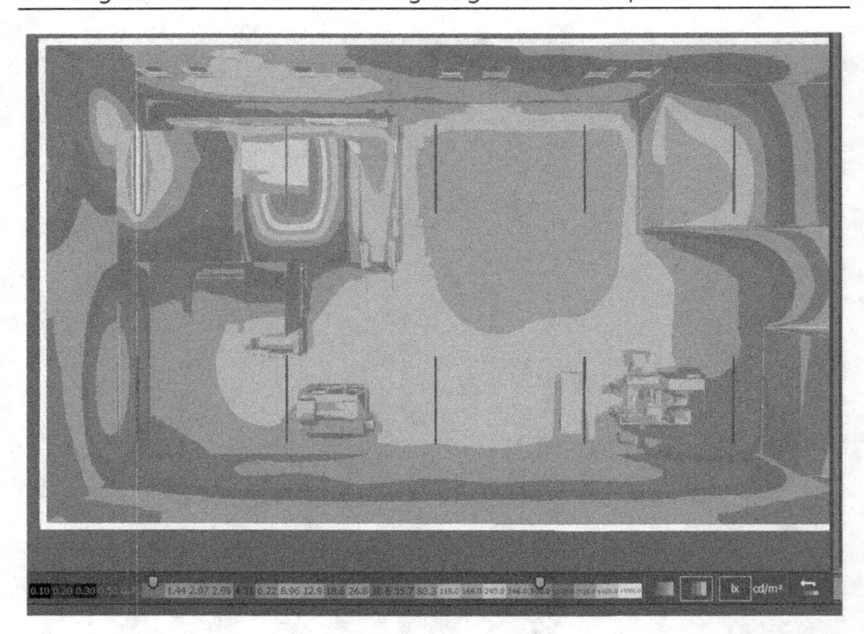

FIGURE 4.11 The course of isophotes of the combined illumination of the production hall's digital twin after rationalization.

TABLE 4.10 Results of illumination intensity of a rationalized solution using a digital twin – combined lighting

CALCULATING AREA	*PARAMETER*	*AVERAGE VALUE [LX]*	*MIN. VALUE [LX]*	*MAX. VALUE [LX]*
Calculating area – Measurement point No. 1	Illumination intensity	421.3	414.2	428.3
Calculating area – Measurement point No. 2	Illumination intensity	490.0	487.5	492.4
Calculating area – Measurement point No. 3	Illumination intensity	504.1	500.6	507.5

Graphical interpretation of the comparison between the in-situ measurements and the rationalization of the intensity of artificial lighting (Figure 4.12) clearly confirms the suitability of the proposed changes for the assessed areas of visual tasks. The illumination intensity at each of the points

TABLE 4.11 Comparison of the results of the current state and ergonomic rationalization of the production hall lighting using digitalization

	ARTIFICIAL LIGHTING		
	MEASURING POINT NO. 1	*MEASURING POINT NO. 2*	*MEASURING POINT NO. 3*
Mean value of in-situ measurement [lx]	147,5	153,6	114,4
Rationalization solution – Lighting intensity [lx]	333	335	355
Deviation – real measurement/ rationalization solution [%]	**+ 125.7**	**+ 118.2**	**+ 210.3**
	DAY LIGHTING		
Mean value of in-situ measurement [lx]	8,4	6,7	7,1
Rationalization solution – Lighting intensity [lx]	20.4	25.8	41.1
Deviation – real measurement/ rationalization solution [%]	**+ 142.3**	**+ 286.1**	**+ 478.2**
	COMBINED LIGHTING		
Mean value of in-situ measurement [lx]	107.6	103.9	152.1
Rationalization solution – Lighting intensity [lx]	421.3	490	504.1
Deviation – real measurement/ rationalization solution [%]	**+ 291.5**	**+ 371.6**	**+ 231.4**

assessed reaches the necessary levels, ensuring appropriate visual conditions primarily for night shift work.

When comparing the intensity of day illumination and the obtained measurement data, it is also possible to confirm the suitability of the proposed solutions. The day illumination intensity parameter is only informative in nature, but it reaches higher values than those obtained by direct measurement, which is subsequently confirmed by the last comparison of the quantitative assessment – the comparison of the combined lighting intensity.

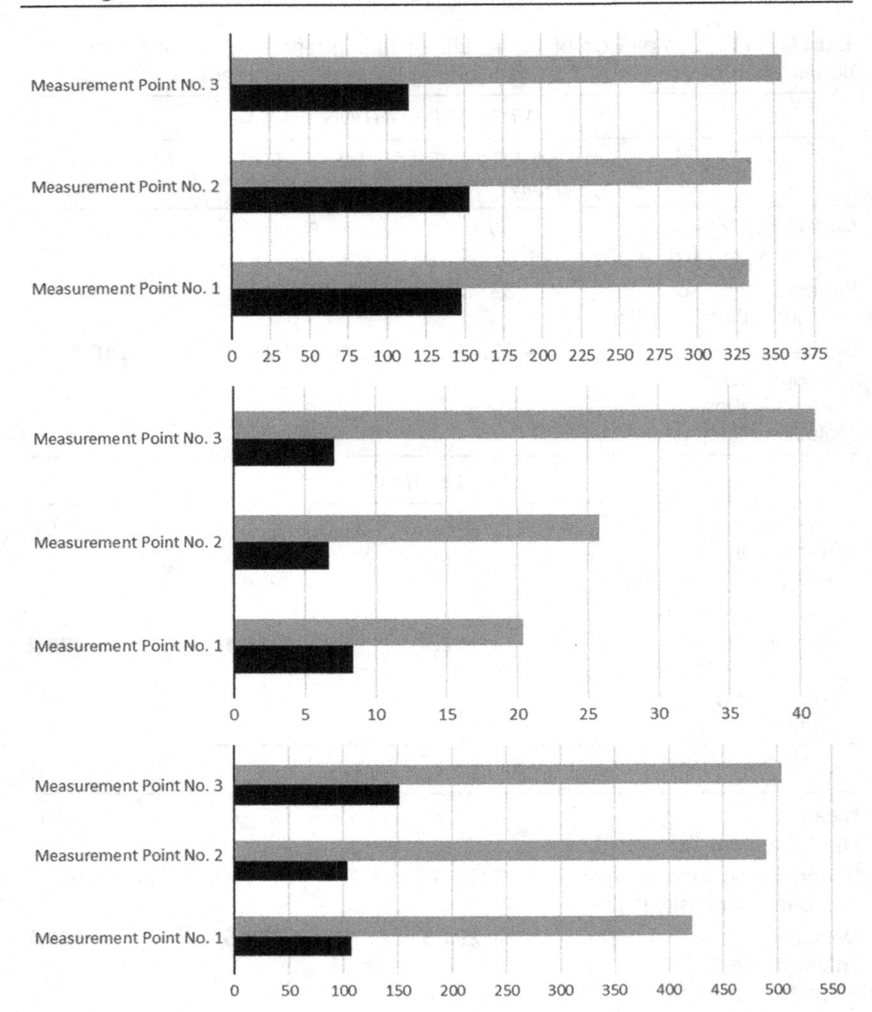

FIGURE 4.12 Graphical interpretation of the comparison of the in-situ measurements and the rationalization of artificial lighting intensity (up), day lighting intensity (middle), combined lighting intensity (down).

Note:

Red interpretation – in-situ measurements
Green interpretation – rationalization solution

Comparison of combined lighting is the most important attribute of the implemented ergonomic rationalization, as employees perform work activities primarily under this type of lighting. According to Figure 4.12, it can also be concluded that in the third case, too, the proposed ergonomic rationalization measure is appropriate, as the values achieved at individual points also exceed three times the values obtained by the in-situ measurement.

In conclusion, it can be stated that the overall assessment of individual illumination parameters achieved at the specified computation levels confirms the complex suitability of the model for real application in practice.

5 Implementation of Digital Ergonomic Transformation in Non-Production Space

The final chapter of the monograph is devoted to the description of the solution of digital ergonomic transformation in another case study realized directly in practice. As an example of this implementation of digital ergonomic tools in the assessment of lighting of non-production areas, an example from practice was selected – the rehabbing of a classroom at the university. The selected classroom underwent a complex reconstruction, during which its current state was assessed with the subsequent implementation of a new artificial lighting design. Like the previous chapter, the introductory part is devoted to the analysis of the analysed room – classroom at the university, where practical in-situ lighting measurements were performed. Based on the analysis, the need for changes to the lighting system was confirmed, which were processed in the first step by ergonomic rationalization using digital tools. This step consisted of the creation of a digital twin of the assessed area and rationalization proposals of model solutions to improve the lighting system. In the final step, the most

DOI: 10.1201/9781003359937-5

rationalizing proposal was analysed from a technical and economic point of view and subsequently implemented in practice.

5.1 IMPLEMENTATION OF IN-SITU MEASUREMENTS

Due to the fact that the aim is a redesign of artificial lighting, the measurements were carried out exclusively during the early morning/late evening hours. The overall dimensions of this classroom are 9 000 × 6 000 × 3 200 (mm). The classroom features four pairs of windows, three rows of benches, a pulpit, and a blackboard. The layout of individual objects with the representation of the measuring points and real layout of room are shown in Figure 5.1.

In the condition before the reconstruction, there were three rows of older lighting fixtures in the classroom, four evenly distributed pieces in each row with the following parameters: lamp input power – 21 W, colour of light 4 000 K, luminous flux 1 720 lm.

Before the measurements were carried out, five measuring points were established, the distribution of which is shown in the previous figure up. The measurements were carried out with the Roline RO-1332 digital luxometer, the same as in the production hall.

As mentioned above, the classroom contains 12 pieces of older light fixtures, which at first glance, due to their age and slightly neglected maintenance, do not meet the lighting requirements set by standards or legislation. The layout of the light fixtures is shown in the previous figure down.

At each of the measuring points shown above, 3 series of measurements were carried out, 10 measurements in each series, the average value of which is recorded in Table 5.1. The last row shows the total average value of individual average values from the measurements carried out for the purpose of subsequent creation of the model in the Dialux evo simulation tool.

In view of the measured values presented in tabular and graphical form above (Figure 5.2), it should be noted that since the value of at least 300 lx as required by the standard and legislation had not been reached, it is necessary to proceed with ergonomic rationalization. To carry out ergonomic rationalization of lighting of the selected space, it is appropriate from the economic

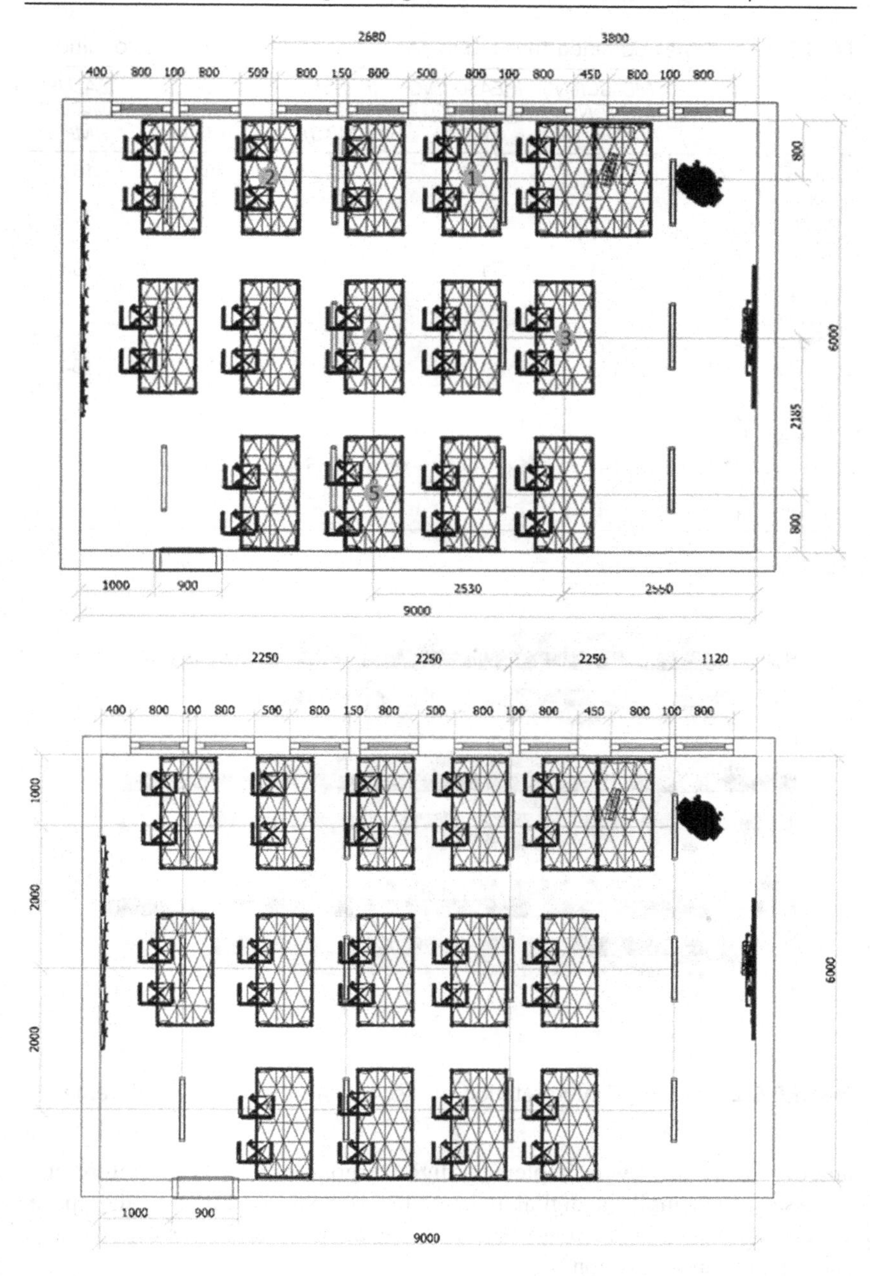

FIGURE 5.1 Classroom floor plan with measuring points (up) and original layout of the lighting system fixtures (down).

TABLE 5.1 Values obtained through in-situ measurement – current condition

	MEASURING POINT NO. 1 (MP 1)	*MEASURING POINT NO. 2 (MP 2)*	*MEASURING POINT NO. 3 (MP 3)*	*MEASURING POINT NO. 4 (MP 4)*	*MEASURING POINT NO. 5 (MP 5)*
1. measurement series of illumination intensity [lx]	138	155	172	184	163
2. measurement series of illumination intensity [lx]	141	157	171	179	168
3. measurement series of illumination intensity [lx]	153	156	177	175	164
Mean value of illumination intensity [lx]	144	156	173.3	179.3	165

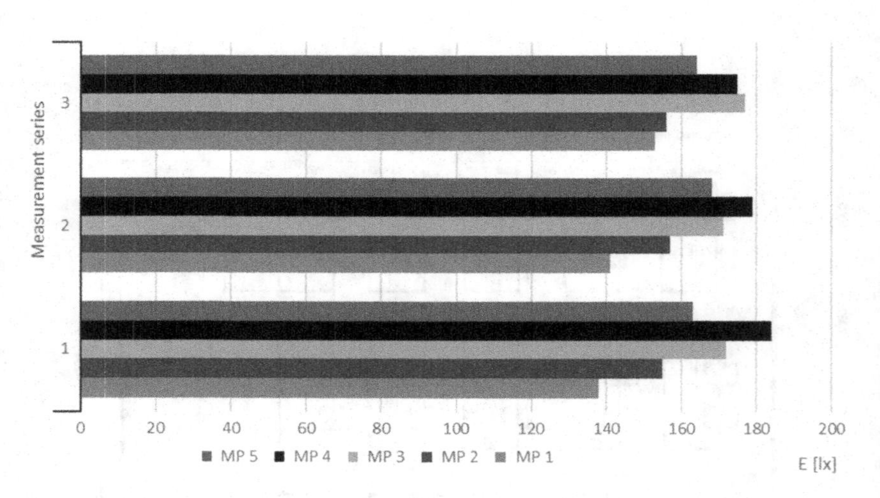

FIGURE 5.2 Graphical interpretation of results from the in-situ measurements.

and time point of view to implement digital ergonomics tools. Therefore, it is necessary to compile a digital twin of the current condition of the space, which can be verified on the basis of the measurements carried out with subsequent rationalization.

5.2 THE USE OF DIGITAL TRANSFORMATION TOOLS IN THE CONSTRUCTION OF A DIGITAL TWIN OF NON-PRODUCTION SPACE

The model of the current condition was compiled according to the real measurement of the positions of individual objects and also on the basis of in-situ measurements. The computation areas are defined in the programme on the basis of the actual measuring points in the environment. Dimensions of each area for computation are 100 × 100 mm and each is situated 0.85 m above the floor. As this is an assessment of artificial lighting, it is also necessary to cater to this condition in the program. It is necessary to turn off the daylight component in the light – light scenes section.

Individual output values were assessed on the basis of the basic work environment setting – in this case, school buildings, classrooms, and teachers' offices. The tool chosen for running the simulation evaluates the results achieved and works as such on the basis of the STS EN 12464–1, CIE 097-2005, and DIN V 18599 standards. The constructed model of the current condition is shown in Figure 5.3. Through this model, the intensity of illumination in a given classroom was subsequently assessed.

FIGURE 5.3 Digital twin of assesses room in current state.

Due to the fact that currently the manufacturers of the light fixtures no longer have the given types of lamps, the TPS682 1xTL5-21W HFP AC-MLO_865 lamp from Phillips with parameters identical to those of the light fixtures currently located in the room was selected from the catalogue for the current condition verification.

Based on the construction of the lighting system, in the specific case of the lighting system for the verification of artificial lighting, an overall model was compiled, and a simulation was carried out, the results of which are shown in Figure 5.4 and the average values are shown in Table 5.2. Subsequently presented is a 3D model with the course of isophotes of the verified lighting of the entire facility.

The results of the analysis of the current condition show that the application of ergonomic rationalization is essential. During the performed measurement and even after the creation of the verification simulation model, the required artificial lighting values were not achieved. The maximum values reached were below 200 lx limit. In view of this situation, digital ergonomic rationalization is further implemented with the construction of two rationalization designs to ensure the required values of artificial lighting are reached.

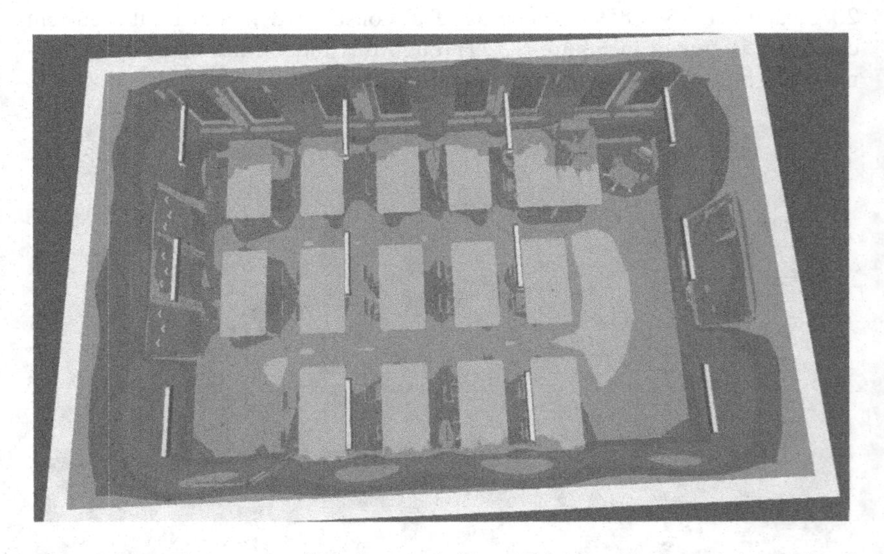

FIGURE 5.4 The course of isophotes of the artificial illumination of the analysed room – current state.

TABLE 5.2 Values obtained through simulation – current condition

	ILLUMINATION INTENSITY		
	MIN VALUE [LX]	*MEAN VALUE [LX]*	*MAX VALUE [LX]*
Measuring point 1 – Calculating area 1	150	152	155
Measuring point 2 – Calculating area 2	157	158	159
Measuring point 3 – Calculating area 3	177	178	180
Measuring point 4 – Calculating area 4	177	179	182
Measuring point 5 – Calculating area 5	161	163	166

5.3 USE OF DIGITAL TRANSFORMATION TOOLS IN THE ERGONOMIC LIGHTING RATIONALIZATION OF NON-PRODUCTION SPACE

The following section describes the origin of two rationalization proposals to improve artificial lighting in the selected workplace in view of the results obtained from the in-situ measurement and the digital twin. In both cases, the same type of lighting was applied and in both cases, it was also assumed the ceiling would be lowered from 3 200 to 2 300 mm.

5.3.1 Replacement of Existing Type of Lighting – The First Design Solution

The first design for achieving the required artificial lighting values in the selected area consists of replacing the existing type of lighting with a new type. As a mass reconstruction is currently taking place in the assessed area and in its surroundings, the design will consider lighting that is to be applied to each renovated classroom. The following text shows the basic parameters of the new lighting fixture:

- Input power: 2 × 36 W
- Colour temperature: 5 208 K
- Luminous flux: 6 700 lm

- Ra: 92
- Operational efficiency: 83.78%

When selecting new lighting, basic parameters such as the installed input power, colour temperature, or luminous flux should always need to be taken into account. It is currently very important to ensure as much daylight as possible in all rooms where any work activity is carried out, but if the work activity is carried out in a given room in the late afternoon or evening hours, in such a case it is necessary to provide artificial lighting that is as close to the nature of the daylight as possible. The provision of artificial lighting of daytime lighting nature is possible by means of an appropriate choice of the colour temperature of the lighting fixture. The colour temperature of artificial lighting should be close to 6 005 K, which is the daylight temperature. In this lighting fixture, the colour temperature is 6 208 K, which is satisfactory and corresponds to approximating the daytime illumination effect.

Similarly, to the verification model, the proposed model also includes five computation areas. The settings of this model are the same as the verification model settings. These are settings of computation surfaces, measuring rasters, work plane settings, etc. Defined just like in the verification model, the computation areas differ only in the definition of their positions.

After compiling a new model, which is based on a verification model and differs only in the type of the light fixture, it is possible to carry out the first simulation and then evaluate the design created. Evaluation of the artificial lighting design is done through considering both the quantitative and qualitative parameters such as intensity, uniformity, stability, safety, colour, brightness distribution, shading, aesthetics, and maintenance.

5.3.1.1 Illumination Intensity

The parameter is of a quantitative nature and can, therefore, be evaluated on the basis of the achieved numerical characteristics. Table 5.3 shows the average values for the illumination achieved in the model proposed as first.

As shown in the previous table, the values of illumination intensities are excessively high. For the given type of room, it is necessary to achieve values higher than 300 lx, but not three times higher. The conclusion from this assessment is that it is possible to reduce the total number of light fixtures. The second rationalization proposal will therefore be to reduce the number of light fixtures. Figure 5.5 also shows the overall course of isophotes after the first rationalization design has been applied.

TABLE 5.3 Assessment of illumination intensity – the first proposed model

	ILLUMINATION INTENSITY		
	MIN VALUE [LX]	*MEAN VALUE [LX]*	*MAX VALUE [LX]*
Measuring point 1 – Calculating area 1	916	924	932
Measuring point 2 – Calculating area 2	912	917	926
Measuring point 3 – Calculating area 3	1044	1048	1052
Measuring point 4 – Calculating area 4	1081	1083	1089
Measuring point 5 – Calculating area 5	1015	1021	1025

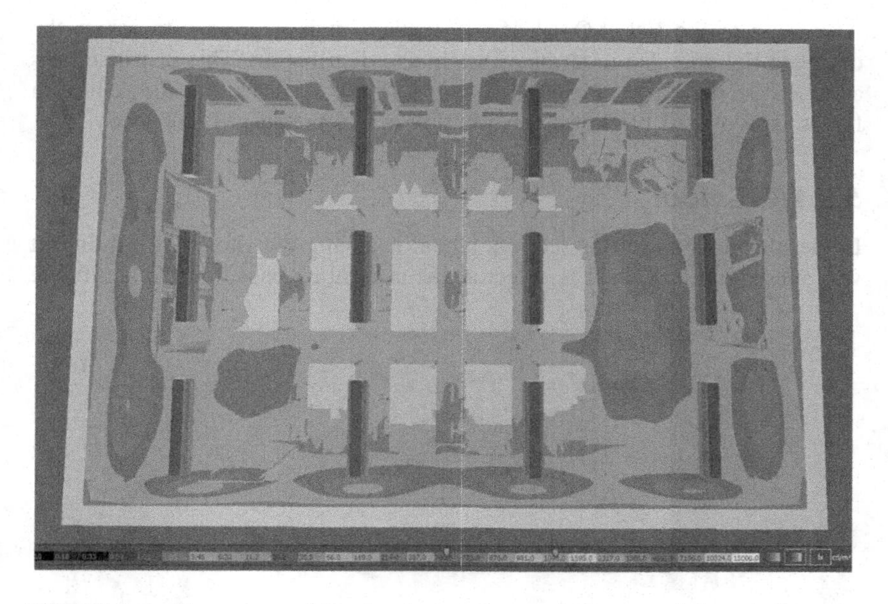

FIGURE 5.5 The course of the isophotes after replacement of the existing type of lighting.

5.3.1.2 Uniformity

Another indispensable parameter to assess is the uniformity of the illumination. In order to ensure light well-being, it is necessary that this quality parameter reaches a value of at least 0.7 or higher. Table 5.4 shows the values achieved by the lighting uniformity parameter.

TABLE 5.4 Assessment of illumination uniformity – first proposed model

	UNIFORMITY [-]
Measuring point 1 – Calculating area 1	0.99
Measuring point 2 – Calculating area 2	0.99
Measuring point 3 – Calculating area 3	1.00
Measuring point 4 – Calculating area 4	1.00
Measuring point 5 – Calculating area 5	0.99

5.3.1.3 Brightness

According to the EN ISO 12 464-1 standard, it is necessary to assess the designed lighting also in terms of brightness distribution. This distribution is assessed on the basis of the illumination of the surfaces and the reflection factor. The overall brightness distribution is interpreted in Figure 5.6.

5.3.1.4 Colour

In the introduction to this section, when justifying the choice of lighting, it was stated that it is ideal to select artificial lighting based on a quantifiable

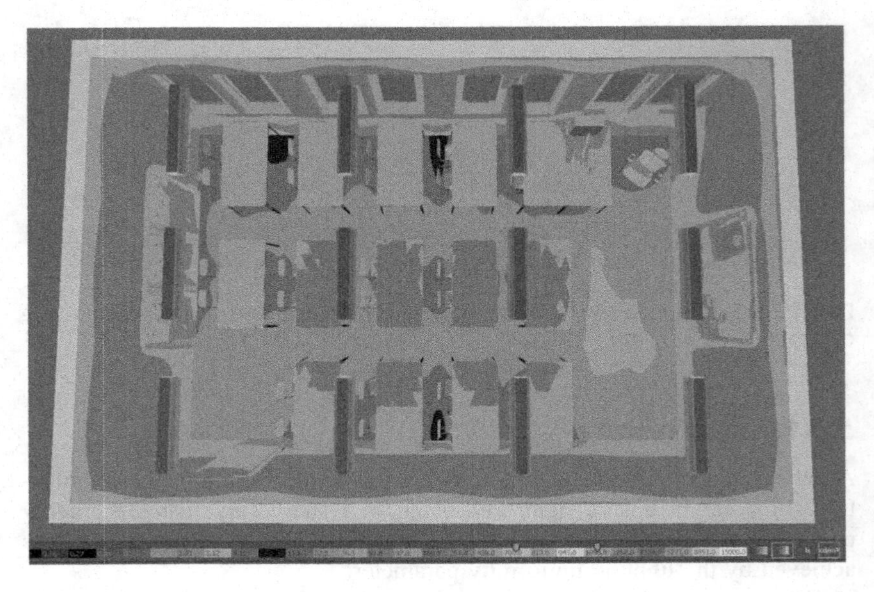

FIGURE 5.6 Illustration of brightness upon replacing the existing type of lighting.

value of the colour temperature so as to approximate this value to daylight. Among other things, this light also ensures light well-being in the work-place and thus good visual performance of the worker. A good indicator of colour evaluation is the (colour rendering index) Ra, which should be at least 80. It is satisfactory in the present case because it reaches a value of up to 92.

5.3.1.5 Maintenance, Aesthetics, and Safety

These three qualitative parameters are assessed in a more or less subjective way. Safety is understood by ensuring the elimination of the stroboscopic effects that occur most often in older light fixtures. As the light fixture is new, this parameter is ensured. Maintenance should be provided at the workplace with a standard factor of 0.8, which has been taken into account in the computations. The last quality parameter is the aesthetics of the lighting, which is met by the choice of a new lighting fixture with an attractive design.

In the overall assessment of the first new proposal, all the required parameters are met, however, the lighting intensity values are unnecessarily high and it is, therefore, advisable not only from a technical but also economic point of view to propose a second solution, which will consist of reducing the number of the light fixtures.

5.3.2 Replacement and Reduction of Existing Type of Lighting – The Second Design Solution

When assessing the illumination intensity of the first artificial lighting design for a given space, it was concluded that it is possible to reduce the number of current light fixtures due to the recorded high values. After a series of experiments and several simulations, five pieces of light fixtures were chosen as the most appropriate number. Figure 5.7 shows their distribution in 2D view, as well as their overall 3D visualization.

Regarding the basic settings of computational objects and all parameters, the composition of the entire system of the constructed model remains the same as that of the system proposed at first. After adjusting the first model to the second one with the reduction of the light fixtures as described above, a description of the achieved results from a qualitative and quantitative point of view followed. Given that the model is only a modification of the previous

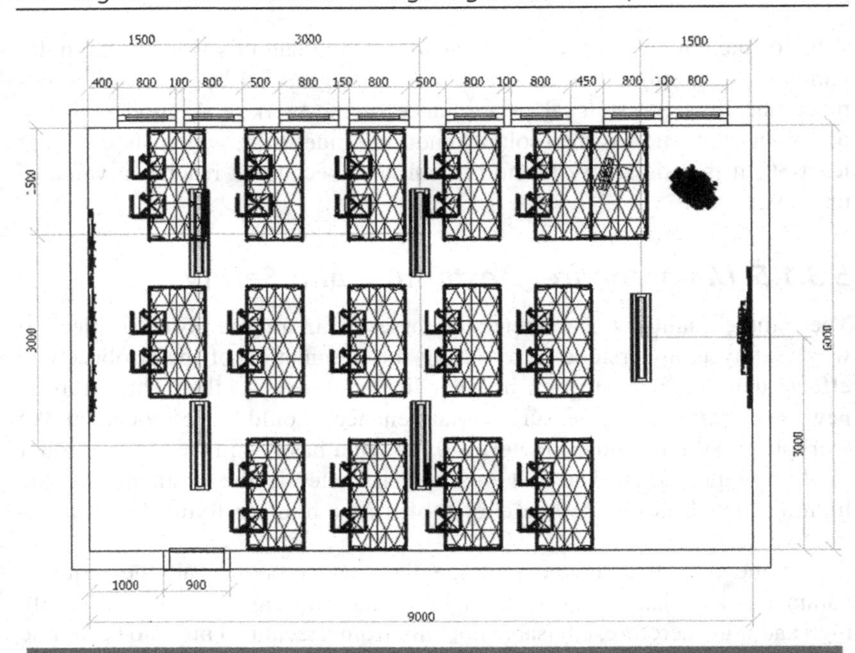

FIGURE 5.7 Layout of the reduced number of light fixtures (up) and post-lighting reduction model (down).

TABLE 5.5 Assessment of illumination intensity – model proposed as second

	ILLUMINATION INTENSITY		
	MIN VALUE [LX]	*MEAN VALUE [LX]*	*MAX VALUE [LX]*
Measuring point 1 – Calculating area 1	399	408	418
Measuring point 2 – Calculating area 2	431	437	443
Measuring point 3 – Calculating area 3	449	451	454
Measuring point 4 – Calculating area 4	476	477	478
Measuring point 5 – Calculating area 5	490	496	501

solution, the qualitative aspect remains unchanged. For the quantitative aspect, the following descriptions are given below.

5.3.2.1 Intensity

Compared to the first proposed solution, it follows from Table 5.5 that the second model also meets all the requirements for average values of illumination intensity as assessed from the average intensity achieved by artificial illumination. At all measuring points, the average value of artificial illumination intensity was recorded in the range of 400–500 lx, which is more than sufficient lighting for the given type of room.

Through the Dialux Evo software, 3D images of the course of artificial lighting isophotes in the plane 0.85 m above the floor were also generated, which can be found in Figure 5.8. The individual values only confirm the results presented in the previous table.

5.3.2.2 Uniformity

Compared to the first proposed solution, there are no significant deviations in the assessment of uniformity (Table 5.6) in the second proposed solution. The difference occurred only at the first measuring point, when the uniformity of lighting decreased by only one hundredth, which is negligible in the present case, as in all cases the uniformity according to EN ISO 12 464-1 with a value of 0.7 or more is observed again.

5.3.2.3 Brightness

According to Figure 5.9, the brightness distribution in this case can also be considered appropriate.

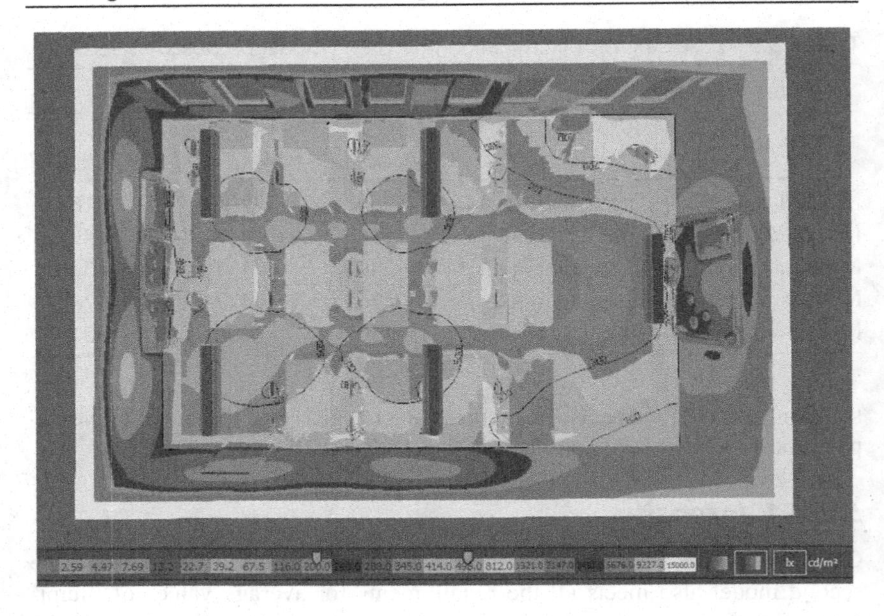

FIGURE 5.8 The course of isophotes under the reduced new type of lighting.

TABLE 5.6 Assessment of illumination uniformity – model proposed as second

	UNIFORMITY [-]
Measuring point 1 – Calculating area 1	0.98
Measuring point 2 – Calculating area 2	0.99
Measuring point 3 – Calculating area 3	1.00
Measuring point 4 – Calculating area 4	1.00
Measuring point 5 – Calculating area 5	0.99

5.3.2.4 Colour, Maintenance, Aesthetics, and Safety

The assessment of these parameters is the same as in the first model proposed, given that the type of illumination remains unchanged. For this reason, it can be concluded that all of these parameters meet the requirements in the same way as was the previous case.

Overall, the second of the proposed solutions can also be considered appropriate, together with the first, but from the economic point of view in particular, it can be considered to be more appropriate.

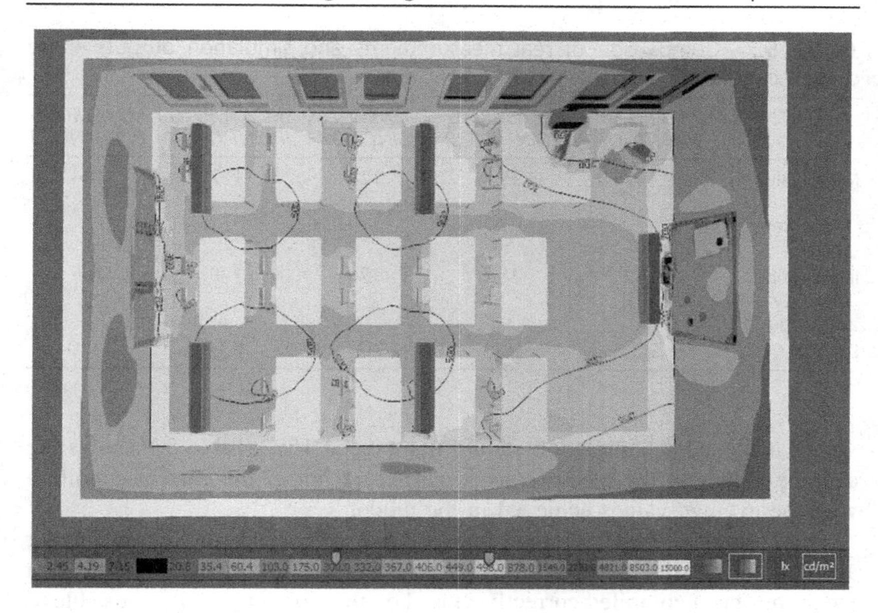

FIGURE 5.9 Illustration of brightness in the new, reduced type of illumination.

The second rationalization solution was directly applied to the assessment room – classroom. The last part of the chapter provides information about the overall assessment of this solution not only in terms of virtual simulation but also in terms of final measurements after the solution has been applied (Figure 5.10).

In the measurements that were carried out during the assessment of the current condition, the overall average value that was necessary to compile the implemented model was determined from the individual average values. According to the available studies, the deviation of the measurement from the

FIGURE 5.10 Implemented rationalization solution.

TABLE 5.7 Comparison of real measurements and simulation outputs – the current condition

	MEASURING POINT NO. 1	*MEASURING POINT NO. 2*	*MEASURING POINT NO. 3*	*MEASURING POINT NO. 4*	*MEASURING POINT NO. 5*
Mean value of in-situ measurement [lx]	147	154	174	177	164
Illumination intensity – simulation [lx]	152	158	178	179	166
Deviation [%]	3.29	2.53	2.24	1.12	1.20

values achieved in simulation should not exceed 15%. Table 5.7 shows the average values of the measurements carried out under the current condition with the average values achieved in the model.

According to Table 5.7 and Figure 5.11, it is clear that at all points under assessment, the deviation below 15% is observed and thus the verification model has been compiled correctly. Based on this correctness, it is possible to subsequently evaluate the proposed solutions.

In the second step, the applied model is evaluated technically – the second proposed solution. This model was selected for practical implementation and after its implementation in the real form, secondary measurements were carried out to confirm the correctness of the model solution. Secondary measurements were carried out using the same measuring instrument under the same conditions as the initial measurements described in the previous chapters. Table 5.8 shows the values from the measurement carried out after reducing the number of light fixtures and lowering the ceiling.

As can be seen from Table 5.8 and the Figure 5.12, the measured values after the application of the second of the proposed solutions are fully compliant with the standards and applicable legislative requirements. In order to

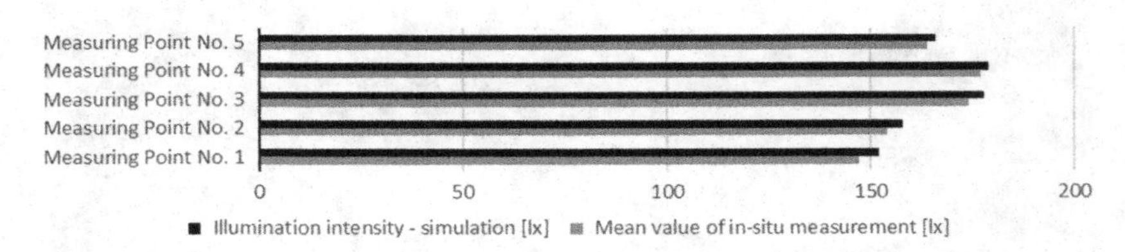

FIGURE 5.11 Graphical interpretation of comparison of real measurements and simulation outputs – the current condition.

TABLE 5.8 Measured values of artificial lighting after application of the proposed solution

	MEASURING POINT NO. 1	*MEASURING POINT NO. 2*	*MEASURING POINT NO. 3*	*MEASURING POINT NO. 4*	*MEASURING POINT NO. 5*
Measurement of illumination intensity no. 1 [lx]	412	430	445	472	485
Measurement of illumination intensity no. 2 [lx]	401	427	452	471	488
Measurement of illumination intensity no. 3 [lx]	405	433	447	473	491
Mean value of illumination intensity [lx]	406	430	448	472	488

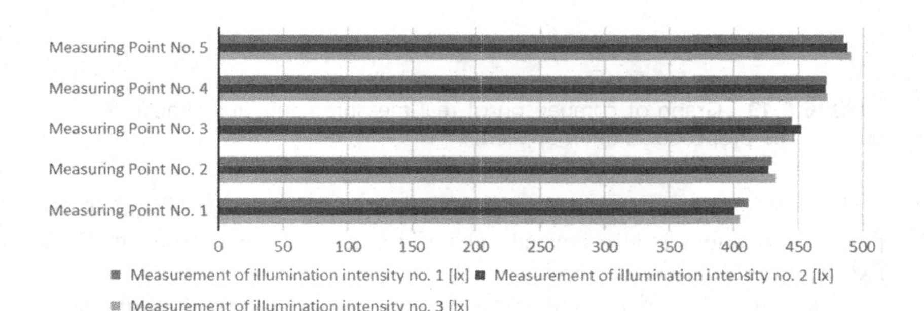

FIGURE 5.12 Graphical interpretation of measured values of artificial lighting after application of the proposed solution.

verify the correctness, a comparison with the compiled model was made similarly to the assessment of the current condition.

As in the assessment of the current condition, when comparing the applied proposed solution and the simulation model, it is necessary to ensure a maximum deviation to be no more than 15%. Table 5.9 and Figure 5.13 provide an overall comparison together with the defined deviation.

As indicated by the values in the table and the figure, in all points the condition for ensuring the deviation between simulation and real measurement not exceeding 15% is met, so it can be concluded that the second proposed and subsequently applied solution is satisfactory.

TABLE 5.9 Table of comparison of real measurements and simulation outputs for the proposed solution applied

	MEASURING POINT NO. 1	*MEASURING POINT NO. 2*	*MEASURING POINT NO. 3*	*MEASURING POINT NO. 4*	*MEASURING POINT NO. 5*
Mean value of in-situ measurement [lx]	406	430	448	472	488
Illumination intensity – simulation [lx]	408	437	451	477	496
Deviation [%]	0.49	1.60	0.67	1.05	1.61

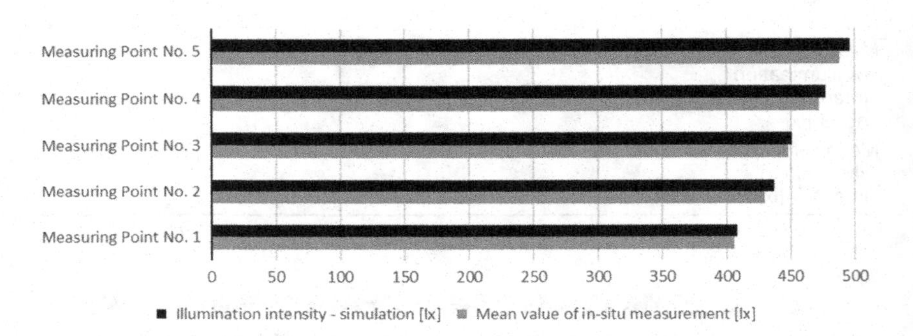

FIGURE 5.13 Graph of comparison of real measurements and simulation outputs for the proposed solution applied.

Due to available options, an economic view of the assessment of the proposed solutions is also created, taking into account the price of the light fixture and fitting work.

The total cost of implementing the first of the proposed solutions is €238.24. Fitting work prices are not included in the total costs, as these works can be carried out by internal sources, and therefore it is not necessary to procure an external company.

Also, in calculating the costs for the second of the proposed solutions, the fitting work is estimated at zero, given that they are carried out by the FVT department. The total costs for both of the proposed solutions are shown in Table 5.10.

On the basis of the economic assessment, it is concluded that the second proposal, namely the replacement and reduction of the lighting system, is the most appropriate alternative of the proposals provided, with the concomitant lowering of the ceiling. This proposal is technically satisfactory and is also a cheaper option by approximately €140.

The return on the abovementioned investments cannot be accurately quantified, given that the total price of health (related to visual impairment) is

TABLE 5.10 Costs of implementing proposed solutions

COSTS OF IMPLEMENTING THE FIRST PROPOSED SOLUTION				
ITEM	*PIECES*	*PRICE PER UNIT (EXCL. VAT)*	*PRICE EXCL. VAT – TOTAL*	*PRICE INCL. VAT – TOTAL*
Mounted light fixture	12	15.88	190.59	238.24
Fitting work	12	–	–	–
Dismantling work	12	–	–	–
Total cost	–	15.88	190.59	**238.24**

COSTS OF IMPLEMENTING THE SECOND PROPOSED SOLUTION				
ITEM	*PIECES*	*PRICE PER UNIT (EXCL. VAT) [€]*	*PRICE EXCL. VAT – TOTAL [€]*	*PRICE INCL. VAT – TOTAL [€]*
Mounted light fixture	5	15.88	79.40	99.25
Fitting work	5	–	–	–
Dismantling work	12	–	–	–
Total cost [€]	–	15.88	79.40	**99.25**

not quantifiable. In addition, the low level of lighting in the room also affects students' concentration. In conclusion, the digital transformation in the concept of ergonomic solutions has its irreplaceable place not only in the assessment of the work environment in production sectors, but also in the ergonomic design of non-production, i.e., learning or office or conventional living spaces.

Conclusion 6

Partial areas of individual sectors are moving at unstoppable speed towards promotion of digital technologies and digitalization as a whole. In the area of production, it is currently possible to identify digitalization not only in the area of production processes themselves, but also in the area of auxiliary and service processes, logistics, storage, or in the continuous provision of comprehensive digitalization of jobs, stations, lines, and halls. The present scientific monograph offers a systematic approach to the field of digital transformation for ergonomics, and after determining comprehensive theoretical knowledge in the field, it describes one of the possibilities of effective use of digitalization also in the field of ergonomics, not only in the production sector when assessing the workspace, but also in the non-production context – ergonomic design of non-production spaces, namely a room intended for educational purposes. The topicality of the scientific monograph is supported by practical solutions, which are shown in the last chapters of the book. The described design solutions are applicable in practice, they can be used in the company and the academic community. The present monograph essentially provides methods of proactive ergonomics, which complements the possibility of digitalization also in this area and thus becomes part of the concept of Industry 4.0 – Digital Manufacturing.

DOI: 10.1201/9781003359937-6

References

Amorim, C. N. D., Cintra, M. S., Sudbrack, L. O., Camolesi, G. E., & Silva, C. (2011). 'Architectural variables and its impact in daylightning: Use of dynamic simulation in Brazilian context'. 27th session of the CIE. South Africa 2011. Proceedings, 376–386.

Ashmore, J., & Richens, P. (2001). Computer simulation in daylight design: a comparison. *Architectural Science Review*, 44(1), 33–44.

Attwood, D. A., Deeb, J. M., & Danz-Reece, M. E. (2004). *Ergonomic solutions for the process industries*. Elsevier.

Badida, M., Králiková, R., & Lumnitzer, E. (2011). Modeling and the use of simulation methods for the design of lighting systems. *Acta Polytechnica Hungarica*, 8(2), 91–102.

Brennan, A., Chugh, J. S., & Kline, T. (2002). Traditional versus open plan office design: a longitudinal field study. *Environment and Behavior*, 34, 279–299.

Bridger, R. (2008). *Introduction to ergonomics*. CRC Press.

Brookes, M. J., & Kaplan, A. (1972). The office environment: space planning and affective behaviour. *Human Factors: The Journal of the Human Factors and Ergonomics Society*, 14, 373–391.

BS EN 12464-1:2021 Light and lighting. Lighting of work places. Indoor work places

Bubb, H., Bengler, K., Grünen, R. E., & Vollrath, M. (Eds.). (2021). *Automotive ergonomics*. Springer Nature.

Bumann, J., & Peter, M. (2019). Action fields of digital transformation – a review and comparative analysis of digital transformation maturity models and frameworks. *Digitalisierung und andere Innovationsformen im Management*, 2, 13–40.

Caterino, M., Manco, P., Rinaldi, M., Macchiaroli, R., & Lambiase, A. (2021l). Ergonomic assessment methods enhanced by IoT and simulation tools. *In Macromolecular Symposia*, 396(1), 1–3.

CIE 13.3 Method of Measuring and Specifying Colour Rendering Properties of Light Sources.

Dado, M., Turek, I., Štetina, J., Bitterer, L., Turek, S., Grolmus, E., & Stibor, P. (1998). *Kapitoly z optiky pre technikov (Chapters in optics for technicians)*. Žilina: Žilinská univerzita.

DIAL GmbH. (n.d.). Dialux software. https://www.dialux.com/en-GB/

Dolníková, E., & Katunský, D. (2017). *Hodnotenie svetelnej pohody pri kombinovanom osvetľovacom systéme v priemyselných halách (Evaluation of light comfort with a combined lighting system in industrial halls)*. Košice: Technická univerzita v Košiciach.

Dolnikova, E., Katunsky, D., & Darula, S. (2020). Assessment of overcast sky daylight conditions in the premises of engineering operations considering two types of skylights. *Building and Environment*, 180, 106976.

Dupláková, D., & Flimel, M. (2017). Layout effect of manufacturing workplace to illumination of working position. *International Journal of Engineering Technologies IJET*, 3(1), 11–13.

Dupláková, D., Hatala, M., Duplák, J., Knapčíková, L., & Radchenko, S. (2019). Illumination simulation of working environment during the testing of cutting materials durability. *Ain Shams Engineering Journal*, 10(1), 161–169.

Flimel, M. (2010). Ergonomic system and management of the light climate on workplaces. *Tome VIII*, 3(3), 277–281.

Flimel, M. (2013). Interactions of the artificial light sources and the window glazing of buildings. *Przegląd Elektrotechniczny*, 89(8), 116–119.

Ghita, M., Cajo Diaz, R. A., Birs, I. R., Copot, D., & Ionescu, C. M. (2022). Ergonomic and economic office light level control. *Energies*, 15(3), 734.

Global lighting association. (2018). Application of CIE 13.3-1995 with Associated CRI-based colour rendition properties. https://www.globallightingassociation.org/images/files/GLA_publication_-_Application_of_CIE_13.3-1995_with_Associated_CRI-based_Colour_Rendition_Properties_-_December_2018_1.pdf

Hao, B., Cheng, M., Li, X., Zhang, Y., Zhao, C., Wang, C., & He, X. (2020). Optimization design and analysis of classroom light grid. Journal of Physics: Conference Series, 1449(1), 1–5.

International Ergonomics Association (IEA). (n.d.). What is ergonomics? https://iea.cc/what-is-ergonomics/

IPC. (n.d.). Ergonomics and work. Why this combination is needed? https://www.ipcworldwide.com/ergonomics-and-work-why-this-combination-is-needed/news/

Kaljun, J., & Dolšak, B. (2012). Ergonomic design knowledge built in the intelligent decision support system. *International Journal of Industrial Ergonomics*, 42(1), 162–171.

Karabulut. C. (2020). *Paperless processes and digitalization in Foreign Trade-Scholar Press*. Mauritius: Scholar´s Press.

Kocisko, M., Teliskova, M., Baron, P., & Zajac, J. (2017). An integrated working environment using advanced augmented reality techniques. In 2017 4th International Conference on Industrial Engineering and Applications (ICIEA), 279–283.

Kralikova, R., & Kevicka, K. (2012). Application of radiosity simulation methods for lighting researches. *New technologies-trends, innovations and research*. IntechOpen.

Králiková, R., Lumnitzer, E., Džuňová, L., & Yehorova, A. (2021). Analysis of the impact of working environment factors on employee's health and wellbeing; workplace lighting design evaluation and improvement. *Sustainability*, 13(16), 8816.

Lali, W. (2021). Connamix – digitization, digitalization or digital transformation? https://connamix.com/digitization-digitalization-or-digital-transformation/

Makaremi, N., Schiavoni, S., Pisello, A. L., & Cotana, F. (2019). Effects of surface reflectance and lighting design strategies on energy consumption and visual comfort. *Indoor and Built Environment*, 28(4), 552–563.

Mandala, A., & Santoso, A. R. (2018). Comparative study of daylighting calculation methods. In SHS Web of Conferences, 41, 1–8.

Nag, P. K. (2019). Lighting systems. *Office Buildings*, 371–404.

Nasir, B. (2018). *Power electronics*. Foundation of technical education.

O'Connor, B., & Secades, C. (2013). *Review of the use of remotely-sensed data for monitoring biodiversity change and tracking progress towards the Aichi biodiversity targets*. UNEP – WCMC.

Panda, A., & Dyadyura, K. (2019). Materials and methods of research. *Monitoring and Analysis of Manufacturing Processes in Automotive Production*, 3, 19–31.

Piňosová, M., Hricová, B., Lumnitzer, E., & Andrejiová, M. (2015). Hodnotenie kombinovaných účinkov rizikových faktorov vo vybranom pracovnom prostredí. (Translation – Assessment of the combined effects of risk factors in a selected working environment). *Fyzikálne Faktory Prostredia*, V, 93–97.

Pokorná, V. (2014). Risk analysis as a tool for improvement of machine operation safety and reliability. *Strojírenská Technológie*, 19(1), 37–42.

Pransky, G. S., & Dempsey, P. G. (2004). Practical aspects of functional capacity evaluations. *Journal of Occupational Rehabilitation*, 14(3), 217–229.

Rashwan, A., El Gizawi, L., & Sheta, S. (2019). Evaluation of the effect of integrating building envelopes with parametric patterns on daylighting performance in office spaces in hot-dry climate. *Alexandria Engineering Journal*, 58(2), 551–557.

Relux Informatik AG. (n.d.). Relux deskop software. https://reluxnet.relux.com/en/

Škoda, J., & Motyčka, M. (2018). Lighting Design Using Ray Tracing. In 2018 VII. Lighting Conference of the Visegrad Countries (Lumen V4), 1–5.

Stachová, K., Stacho, Z., Cagáňová, D., & Stareček, A. (2020). Use of digital technologies for intensifying knowledge sharing. *Applied Sciences*, 10(12), 4281.

Trebuna, P., Petriková, A., & Pekarcikova, M. (2017). Influence of physical factors of working environment on worker's performance from ergonomic point of view. *Acta Simulatio*, 3(3), 1–9.

Waterson, P., & Eason, K. (2009). '1966 and all that': trends and developments in UK ergonomics during the 1960s. *Ergonomics*, 52(11), 1323–1341.

What are digital twins? Three real-world examples. (2022). https://daystech.org/what-are-digital-twins-three-real-world-examples/

Wilson, J. R. (2000). Fundamentals of ergonomics in theory and practice. *Applied Ergonomics*, 31(6), 557–567.

Index

Printed in the United States
by Baker & Taylor Publisher Services